SOILS
AND
ENVIRONMENTAL
QUALITY

GARY M. PIERZYNSKI
DEPARTMENT OF AGRONOMY
KANSAS STATE UNIVERSITY
MANHATTAN, KANSAS

J. THOMAS SIMS
DEPARTMENT OF PLANT AND SOIL SCIENCE
UNIVERSITY OF DELAWARE
NEWARK, DELAWARE

GEORGE F. VANCE
DEPARTMENT OF PLANT, SOIL AND INSECT SCIENCES
UNIVERSITY OF WYOMING
LARAMIE, WYOMING

LEWIS PUBLISHERS
Boca Raton   Ann Arbor   London   Tokyo

Library of Congress Cataloging-in-Publication Data

Pierzynski, Gary M.
    Soils and environmental quality/by Gary M. Pierzynski, J. Thomas
Sims, and George F. Vance.
      p. cm.
    Includes bibliographical references (p.   ) and index.
    ISBN 0-87371-680-9
    1. Soil pollution. 2. Soil remediation. 3. Soils. I. Sims, J.
T. II. Vance, George F. III. Title.
TD878.P54 1993
628.5′5 — dc20

93-23049
CIP

© 1994 by CRC Press, Inc.
Lewis Publishers is an imprint of CRC Press

No claim to original U.S. Government works
International Standard Book Number 0-87371-680-9
Library of Congress Card Number 93-23049
Printed in the United States of America   3 4 5 6 7 8 9 0
Printed on acid-free paper

Dedicated to the families of the authors

Garrison, Jeanne, and Joy
Connie, Amy, Ethan, and Katy
Christy, Emily, and Maureen

whose love, patience, and support made this book a reality.

The quality of our environment is a function of both natural phenomena and human activities. Soils are often the interface between human activities and those parts of the environment that we wish to preserve and protect. An understanding of soil properties and reactions is therefore critical to the evaluation of how many contaminants, as well as essential nutrients, behave in the soil environment. Soils play an important role in environmental quality as they can be a source, sink, or interacting medium for many contaminants. Consequently, proper soil management is an important step in maintaining and even improving environmental quality.

In this book we first provide an overview of basic soil science, hydrology, atmospheric chemistry, and the classification of pollutants. This is followed by comprehensive discussions of the role of soils in the biogeochemical cycling of major elements and compounds of environmental concern. Nitrogen, phosphorus, sulfur, trace elements, organic chemicals, greenhouse gases, and acid precipitation are discussed in depth. Interactions of these potential pollutants with soils and the aquatic and atmospheric environments are emphasized. Methods of soil management or remediation to minimize or correct pollution are presented. The concept of risk assessment is reviewed using several contemporary examples, such as pesticide concentrations in drinking water and contamination of soils by trace elements in organic wastes.

Soils and Environmental Quality was written for use as a text for an upper level undergraduate course in soils and the environment. The book was written at a level that requires the reader to have a basic knowledge of chemistry. With appropriate supplementation, we believe that the book could be used as a basis for a graduate level course as well. Individuals with an interests in soils, environmental engineering, forestry, environmental science, biology, geology, or geography will also find this book useful. As readers of this book, we would appreciate your comments and suggestions that you feel would assist us in improving future editions. We are especially interested in student and instructor views on how beneficial the book was, either as a primary or secondary text.

Gary M. Pierzynski
Department of Agronomy
Throckmorton Hall
Kansas State University
Manhattan, KS 66506-5501

J. Thomas Sims
Department of Plant and
   Soil Science
Delaware Ag. Exp. Station
University of Delaware
Newark, DE 19711-1303

George F. Vance
Department of Plant, Soil,
   and Insect Sciences
P.O. Box 3354
University of Wyoming
Laramie, WY 82071-3354

ACKNOWLEDGMENTS

The authors would like to acknowledge Drs. S. A. Boyd, M. J. Brewer, R. L. Chaney, M. B. David, J. M. Ham, R. J. Lavigne, D. E. Legg, T. J. Logan, S. D. Miller, L. C. Munn, C. E. Owensby, A. L. Page, J. A. Ryan, A. Sharpley, S. Sprecher, B. Vasilas, and S. E. Williams for their reviews of portions of this book. Their comments were extremely valuable and helped improve this book immeasurably. The authors would also like to thank Tim Brewer, Barbara Broge, and Susan Glaze for their art work on numerous figures.

Gary M. Pierzynski is currently an Assistant Professor of Soil Chemistry and Fertility in the Department of Agronomy at Kansas State University, Manhattan, Kansas. He received his B.S. degree in Crop and Soil Science (1982) and M.S. degree in Soil Environmental Chemistry (1985) from the Department of Crop and Soil Sciences at Michigan State University, East Lansing, Michigan. He earned his Ph.D. degree in Soil Chemistry (1989) from the Department of Agronomy at The Ohio State University, Columbus, Ohio.

Dr. Pierzynski's research interests include trace element chemistry, remediation of trace element contaminated soils, phosphorus bioavailability, phosphorus and sulfur nutrition for various crops, land disposal of waste materials, and nitrogen fertilizer use efficiency. He has been active in a national committee of the Soil Conservation Service whose charge has been to develop an index for assessing various landforms and management practices for potential risk of phosphorus movement to water bodies. He teaches Environmental Quality, the outline of which served as the starting point for this book, and Plant Nutrient Sources. He is a Natural Resources and Environmental Sciences faculty member at Kansas State University.

Dr. Pierzynski would like to acknowledge two mentors, Dr. Lee W. Jacobs and Dr. Terry J. Logan, who provided superb guidance for two graduate degrees and who continue to serve as sources of sound advice, and the late George Landsfeld, who had faith and would not allow the dream of a doctoral degree to go unfulfilled.

J. Thomas (Tom) Sims is a Professor of Soil Science in the Department of Plant and Soil Sciences at the University of Delaware, Newark, Delaware. He received his B.S. degree in Agronomy (1976) and M.S. degree in Soil Fertility (1979) from the Department of Agronomy at the University of Georgia. He earned his Ph.D. degree in Soil Chemistry (1982) at the Department of Crop and Soil Sciences, Michigan State University, East Lansing, Michigan.

Dr. Sims teaches courses in *Environmental Soil Management* and *Soil Fertility and Plant Nutrition* (undergraduate and graduate) and conducts an active research program directed towards many of the environmental issues faced by agriculture in the rapidly urbanizing northeastern United States. His research has focused on the development of nitrogen and phosphorus management programs that maximize crop yields while minimizing the environmental impact of fertilizers and animal manures on ground and surface waters. Other research has evaluated the potential use of sludge composts, coal ash, and other industrial by-products as soil amendments. Again, the goal has been to develop environmentally sound management programs for these materials, based on their reactions in the soil and effects on plant growth and water quality. In his role as director of the University of Delaware Soil Testing Program, he has developed and evaluated soil tests for environmental purposes such as soil nitrate testing, environmental soil tests and field rating systems for phosphorus, and soil testing strategies for heavy metals in waste-amended soils.

 **George F. Vance** is currently an Assistant Professor of Soil and Environmental Chemistry and Head of the Soil Science Section in the Department of Plant, Soil and Insect Sciences at the University of Wyoming, Laramie, Wyoming. He received his B.S. degree in Crop and Soil Sciences (1981) and M.S. degree in Soil Pedology/Soil Organic Chemistry (1985) from the Department of Crop and Soil Sciences at Michigan State University, East Lansing, Michigan. He earned his Ph.D. degree in Soil Chemistry (1990) from the Department of Agronomy and was a Postdoctoral Research Scientist in the Department of Forestry both at the University of Illinois, Urbana, Illinois.

Dr. Vance's research interests include the chemistry of selenium in range and mine land ecosystems, groundwater contamination by inorganic and organic pollutants, forest nutrient cycling with particular emphasis on organic nutrient forms, chemistry of waste constituents in the soil-plant continuum, and pesticide mobility and fate in semi-arid and arid environments. He teaches a variety of courses at the University of Wyoming including Soil and Environmental Quality, Chemistry of the Soil Environment, Analytical Methods for Ecosystems Research, and Soil Organic Chemistry. He has been actively assisting in the development of a new Agroecology undergraduate curriculum within the Plant, Soil and Insect Sciences Department and in the implementation of the university-wide Environmental and Natural Resources option program at the University of Wyoming. While at the University of Illinois, Dr. Vance was recognized as one of the outstanding teachers of the University (1986) for his student teaching in Introductory Soils.

TABLE OF CONTENTS

SOILS
AND
ENVIRONMENTAL
QUALITY

1   INTRODUCTION TO ENVIRONMENTAL QUALITY

1.1 INTRODUCTION

The general topic of environmental quality can be approached from many different perspectives. Engineers, geographers, architects, lawyers, business people, health care specialists, biologists, and geologists, to name a few, could all be concerned with environmental quality. This book takes a soil scientist's approach to environmental quality. Soil scientists have a unique vantage point for environmental quality because numerous materials that can be considered pollutants cycle through soil. Many of the inputs used in production agriculture can be pollutants, and soils are often inadvertently contaminated with various substances not associated with production agriculture. Soil itself can be a pollutant. Soils also interact with the hydrologic cycle and the atmosphere by serving as a source or sink for various constituents in water and air.

To better understand the relationship between the production of food and environmental quality, consider the manufacture of the fictitious product called the "widget" that students of business frequently use as a learning aid. If the production of widgets resulted in an unacceptable risk to the environment, then society could decide to get along without widgets. The same cannot be said of food. We must continue to produce food, and therefore society must take responsibility for the potentially hazardous materials used in production agriculture. Note the use of the word society. Ideally, all individuals should understand the basic food production processes including the risk versus benefit issues associated with the inputs.

The work of soil scientists is not confined to boundaries imposed by agronomy; however, because soil scientists are concerned with both rural and urban issues alike. Outside the arena of food production, we must realize that as long as hazardous materials are stored, transported, utilized, or produced as by-products, the chance exists that there will be accidental releases into the environment. A risk-free society cannot exist. In the context of this book, soils are frequently the recipients of many materials not associated with production agriculture, either accidently or intentionally. The result may be contaminated soils. In all likelihood, we are not yet aware of the extent of the problem of contaminated soils. Contaminated soils will continue to be discovered and, unfortunately, will continue to occur.

1.2 ENVIRONMENTALISM

It may seem that environmental quality is a new topic when, in fact, the current emphasis on environmental quality is a renewed interest in an old topic. Modern day environmentalism is approximately 30 years old if one considers the publication of *Silent Spring* by Rachel Carson in 1962 the beginning of large-scale environmental awareness by the general public. The U.S. Environmental Protection Agency (EPA) is nearly 20 years old as are a number of scientific journals having the sole purpose of reporting on environmental research. The 20th anniversary of Earth Day occurred in 1990. Two factors are likely responsible for this renewed interest. First, we are becoming increasingly aware that exposure to various materials may reduce our longevity. The natural inclination is to eliminate or reduce our contact with materials that we suspect may adversely affect human health. Second, expanded analytical capabilities, both in terms of what can be detected and at what detection limits, are allowing a more accurate inventory of our environment.

A more philosophical approach would place an individual's attitudes about an environmental issue into one of three categories: *egocentric,* in which an individual's actions are guided solely by concern for him/herself; *homocentric,* meaning concern for the human species; and *ecocentric,* meaning an overall concern for the environment. With each cycle of interest in environmental quality individuals likely progress slowly from egocentric attitudes toward more ecocentric attitudes.

Environmentalism has been a social movement. Eric Hoffer, in *The True Believer,* identified three stages of social movements. The first stage is called *people of words* (updated to be gender neutral). In this stage the social movement is discussed and debated within a small segment of society generally described as the intellectuals. Late in the first stage the social movement may be described in various writings. Rachel Carson was a person of words. The second stage is called *fanaticism* and is characterized by an increase in society's awareness of the movement through various attention grabbing, often illegal, actions performed by zealous believers. Environmental fanaticism may include various means of disrupting the production (e.g., vandalism) of an industry that is believed to be polluting the environment or even something as drastic as bombings or kidnappings. The final stage is called *practical people of action* or, by some, *institutionalism.* At this stage there is widespread awareness and support for the movement. Protests utilize legal means and the potential for violence decreases. Organized boycotts may be employed, for example. Committees or even agencies may be formed. Ecocentric attitudes would prevail in the context of environmentalism. Eric Hoffer (1951) summarizes by stating that "a movement is pioneered by men of words, materialized by fanatics, and consolidated by men of action."

Environmentalism as a whole has clearly progressed to the institutional stage with the federal Environmental Protection Agency and comparable units of government in each state. Within environmentalism there will continue to be issues that arise that will pass through some or all stages of a social movement. Recent attempts by members of environmental groups to physically block the passage of whaling ships would be an example of fanaticism. The issue of global climate change due to increases in the atmospheric concentrations of the greenhouse gases is late in the people of words stage with no indication of fanaticism as yet.

1.3 STUDYING THE ENVIRONMENT: THE SCIENTIFIC METHOD

The *scientific method* provides a set of rules by which the scientific community conducts investigations so that experimental design, collection of data, and interpretation of data are done in a systematic and objective fashion. There is no agreed upon definition of the scientific method in the sense that the same steps are followed in every scientific investigation. There are some common factors in all investigations, however, that must be present before a particular study is acceptable to the scientific community.

A simple deductive process can be used to explain the scientific method. The first step is an *observation*. An observation could simply be poor plant growth or elevated concentration of a heavy metal in a soil sample. Observations can be made with any of the five senses or with instrumentation. Scientific instruments merely enhance our senses. Observations must be verifiable. Other investigators must be able to reproduce the conditions associated with the observation and get the same effect. If alternate techniques and tests are used in the verification, it adds more validity to the observation. Observations that stand the test of verification become accepted as scientific facts.

A considerable amount of science is involved with making and verifying observations and much useful information comes from this. Simple observations, however, do not explain mechanisms, that is, what caused the effect that was observed. For example, the application of coal fly ash to soil at high enough rates will inhibit plant growth. That is a verifiable observation. Coal fly ash can alter soil pH and contains boron and soluble salts, any of which can inhibit plant growth. The second step in the scientific method is the *formulation of hypotheses* and the *testing of hypotheses with controlled experiments*. A *hypothesis* is a plausible explanation for the observation that is tentatively accepted and may serve as a basis for further investigation. *Controlled experiments* are replicated experiments in which all factors except the one in question are held constant. "The boron in the fly ash inhibits the growth of plants when fly ash is applied to soil" is a reasonable hypothesis. A suitable experiment could be designed to determine whether the hypothesis is true or false. Progress would be made regardless of the outcome of the experiment since we would know if boron inhibited plant growth or something else did. If the hypothesis were true, the next logical step would be to ask why boron inhibited plant growth. The process continues.

The replication of treatments in controlled experiments is an important point. Replications provide a means of determining whether treatment effects are real (statistically significant) or due to chance (not statistically significant). The more replications that are used, the more certain the investigator can be that treatment effects did or did not occur.

Certain types of scientific investigations do not lend themselves to controlled experiments. Human health studies are a perfect example. Physicians may observe that patients with a certain type of cancer have been exposed to a particular chemical, but they cannot test the appropriate hypothesis with controlled experiments using human subjects. Epidemiological studies may be useful in these situations. *Epidemiology* is the study of the occurrence and nonoccurrence of disease in a population without the benefit of controlled experiments. If comparable groups (similar distributions of age, sex, etc.) that have been exposed to various levels of the chemical in question can be identified, then statistical comparisons of the frequency of diseases between the groups can be made. Epidemiological studies are not as good as controlled experiments, and this can lead to some interesting statements. Tobacco companies have defended themselves with the argument that no one has ever conclusively proved that the use of tobacco products causes cancer. Technically they are correct because there have been no controlled experiments with humans. Health officials cite the epidemiological studies that show a strong association between tobacco use and the incidence of cancer.

Most scientific investigations end at this point. Occasionally a theory is proposed that attempts to explain or predict a large number of observations or facts — the theory of evolution, for example. Theories are also tested with hypotheses. Then there are natural laws and basic principles. Exceptions to natural laws and basic principles have never been observed. The law of gravity or the conservation of energy are examples.

Publication of the results of scientific investigations in refereed journals is an integral part of the scientific method. The process begins with the submission of a manuscript to a journal editor who then sends it out to several reviewers. The reviewers evaluate the work for uniqueness, adherence to the scientific method, statistical analysis, quality of writing, and overall scientific vigor before it is accepted for publication in the journal. Then the manuscript appears in the journal; and all readers are free to accept, challenge, or attempt to verify the findings. Recall the recent interest in the idea of cold fusion. In this instance, the results of an experiment were released prior to the work going through the publication process. A great deal of media and government attention was generated because of the far-reaching ramifications of cold fusion. When the details of the experiment were published, no one was able to verify the results and a great deal of embarrassment was realized by the investigators and their institution.

The importance of the scientific method cannot be overstated. Environmental issues can be emotionally charged because they can influence human or animal health and involve large sums of money. Decisions or actions based on poor information can result in wasted effort, wasted money, and needless regulations. Being objective requires that you evaluate information. Recognize the source of information and whether it is a product of the scientific method. Recognize information that may be biased because of vested or emotional interests. The scientific method provides a mechanism by which the obtainment of knowledge can proceed in an unbiased fashion.

1.4 ENVIRONMENTAL SCIENCE AND THE GENERAL PUBLIC

Responsibility for the environment ultimately belongs to society. Society elects the people who promulgate the environmental regulations and who decide how tax dollars will be spent on environmental research. The responsibility is not straightforward, however, because many complicated interactions exist. Environmental regulations directly impact industries that provide jobs for society. The costs of compliance with environmental regulations by private businesses and governmental units themselves are passed on to consumers through increased prices and taxes. These interactions would tend to make for weaker environmental regulations. Countering this tendency are peoples' fears of the effects of pollutants on themselves and growing respect for the environment in which they live.

The media plays a role by transferring knowledge from the scientific community to the general public. If the scientific community was an organized entity with its members always in agreement that only released information that had been rigorously evaluated, then the public would be assured of receiving only the most reliable information available. This is not the case, of course. The media reports information that may be of interest to its audience provided the information comes from what the media perceives as a credible source. Little consideration is given to whether the information is a product of a portion or all of the scientific method. Be aware of the difference between an *environmental event* and an *environmental issue*. An environmental event is an important occurrence related to the environment while an environmental issue is a point or question related to the environment that is to be debated or decided. A chemical spill is an environmental event while the response to the spill, which might include actions that need to be taken to prevent additional spills, is an environmental issue. Reporting of issues and events ought to be objective, but objectivity is compromised most often when it comes to issues. Once again, the ability to evaluate information is the key to a complete understanding of environmental concerns.

In addition to the technical aspects involved in studying the relationship of soils to environmental quality, a topic of increasing importance is that of *risk assessment*. *Risk* is the chance of injury, loss, or damage. In the context of environmental science, risk assessment is the process used to quantitatively estimate the risks associated with exposure of any organism to various substances in the environment. Risk assessment can provide the basis for environmental regulations, although the information can be ignored if society chooses to do so.

1.5 SUMMARY

As one studies environmental science, part of the task is becoming familiar with the technical aspects, such as the scientific method, nomenclature, and processes; and part of the task is appreciating the objectivity, philosophical approaches, and even moral questions required to have a complete understanding of environmental issues. Soils play a major role in the cycling of many environmental contaminants, and soil science serves as a useful discipline from which to study the environment. To do so one must have a basic understanding of soils, hydrology, and the atmosphere, which this book will attempt to provide. The major classes of soil pollutants will then be identified followed by a detailed discussion of each. Risk assessment, as related to soils, comprises the final chapter.

REFERENCES

Hoffer, Eric, *The True Believer. Thoughts on the Nature of Mass Movements,* Harper & Row, New York, 1951.

SUPPLEMENTARY READING

Rousseau, D. L., Case studies in pathological science, *Am. Sci.,* 80, 54, 1992.

2

OUR ENVIRONMENT: ATMOSPHERE, HYDROSPHERE, SOILS

2.1 INTRODUCTION

Our environment is comprised of natural wonders that provide for the ingredients of life. Life as we know it requires nutrients, water, and oxygen to survive. These staples of life are provided through the air we breath, the fluids we drink, and the foods we eat. The atmosphere contains essential gases such as O_2, CO_2, and N_2, that are needed to sustain our existence. The water we drink comes from surface- and groundwater supplies, all of which have been cycled over the millions of years the earth has existed. Production of crops for human and animal consumption relies on our ability to plow, seed, and cultivate our lands.

Throughout this book we will be discussing the importance of environmental quality as it relates to the health and well-being of plants, animals, and humans, as well as aesthetic conditions. When we discuss the essential and toxic properties of various elements and chemicals, it is important to recognize the environmental conditions under which they are being considered. A concept that is critical to this relationship is that of nutrient and toxin *bioavailability*. Bioavailability is the possibility that a chemical in the environment will cause an effect, positive or negative, to a specific organism. Bioavailability is a function of the chemical species present, the ability of the organism to readily absorb or ingest it, and the consequences the chemical species has on the organism. Another concept that is worth mentioning at this time is that of *bioaccumulation*. Bioaccumulation is the process in which an organism accumulates an element or substance once it has been absorbed or ingested. In some cases, an organism may accumulate an element at three, four, or five orders of magnitude the concentration found in the environment in which it lives.

This chapter will examine the salient properties of atmosphere, hydrosphere, and soil components that make up our environment. An understanding of these systems is critical to the evaluation of the quality of our environment.

2.2 THE ATMOSPHERE

We live in a time where concern for atmospheric quality is growing due to the increasing amounts of pollutants that are being added to the atmosphere daily.

Atmosphere quality is as important as soil and water quality issues, since the atmosphere surrounds the earth, whereas the soil and hydrosphere components do not. For this reason, the atmosphere plays an important role in nutrient and contaminant transport processes.

The atmosphere consists of several layers (*troposphere, stratosphere, mesosphere,* and *thermosphere*), each with distinct properties. Of the four major layers, only the troposphere and stratosphere, which comprise the lower atmosphere (0–50 km), will be examined due to the importance of the troposphere in biogeochemical processes and the role of the stratosphere in global transport of some materials and ozone chemistry.

Temperature and pressure variations in the lower atmosphere are shown in Figure 2-1. Atmospheric pressure is an approximate logarithmic function of altitude and decreases with height above sea level. Air temperature, however, varies considerably with time of day and season, latitude, and altitude. Atmospheric heat comes from the radiation of heat reflected and emitted from the earth's surface. This is why the temperature of the troposphere decreases with altitude, whereas the adsorption of ultraviolet light by ozone in the stratosphere causes an increase in temperature with altitude. A transition layer between the troposphere and stratosphere is called the tropopause. The tropopause is a region of cold temperature that acts as a protective barrier to the loss of water to the stratosphere, where it can be decomposed.

Chemical composition of the troposphere is relatively homogeneous except in areas which are affected by air pollution. Mixing takes place in the troposphere due to winds and rising of warm air that develop near the surface of the earth. The stratosphere remains relatively unchanged over time since very little vertical mixing takes place.

Humans have had a profound affect on the composition of the atmosphere. Some of the trace gases that have increased due to human activities include CO_2, CH_4, N_2O, CO, and chlorofluorocarbons (CFCs). These gases can either directly or indirectly increase the absorption of infrared radiation, thus intensifying the greenhouse effect (see Chapter 10 for further discussion). Although it may not be readily apparent at the present time, significant climate change may result from pollutant gases and particulates that are entering the atmosphere from burning fossil fuels, removal of natural vegetation, release of CFCs, and other human activities.

2.2.1 Nature and Properties of the Lower Atmosphere

The chemical components making up the majority of the troposphere gases are N_2 and O_2, with Ar, CO_2, and H_2O as major secondary components. Ozone (O_3), although only minor in abundance, is particularly important to the chemistry of the earth's atmosphere. A detailed list of the average global composition of the atmosphere is given in Table 2-1.

Due to the magnitude of N_2 (78.08%) and O_2 (20.95%) in the atmosphere, these elements remain relatively constant over time. The percentage of other atmospheric constituents may change with time as a result of increased natural and anthropogenic emissions. Atmospheric pollutants tend to be higher over source areas such as cities and fossil fuel burning power plants. However, there are also natural sources that

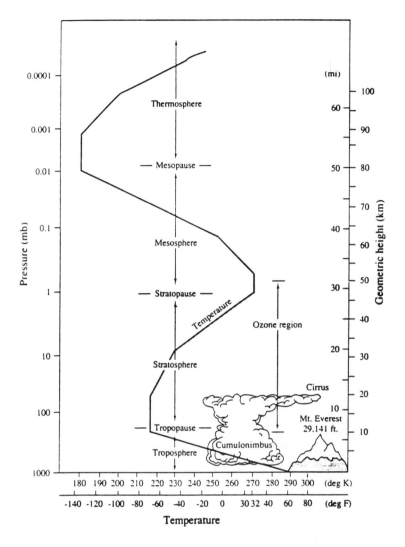

Figure 2-1 Variations in pressure and temperature in the different atmospheric zones. (From Schlesinger, 1991. With permission.)

emit gases: swamps and anaerobic environments release methane (CH_4) and hydrogen sulfide (H_2S); and wildfires can cause the emission of CO_2, CO, CH_4, and carbonyl sulfide (COS). Agricultural practices are also partially responsible for the release of CO and N_2O from the cultivation and fertilization of soils. Even domestic ruminant animals contribute approximately 60–100 metric tons of CH_4 per year, which accounts for 15% of the CH_4 emitted globally.

The atmospheric CO_2 concentration has increased by about 25% over levels determined from preindustrial times, around 1800. Between 1958 and 1988, the level of atmospheric CO_2 rose from 316 to 350 ppm, a rate of increase of over 1 ppm/year. Under the scenario of "business as usual", it has been predicted that the CO_2

Table 2-1   Global Average for the Chemical Composition of the Atmosphere

Chemical constituent	Common name	Percent	Approximate mass (kg)
	Dry air	100	5.12×10^{18}
N_2	Nitrogen	78.08	3.87×10^{18}
O_2	Oxygen	20.95	1.18×10^{18}
Ar	Argon	0.93	6.59×10^{16}
H_2O	Water vapor	0.1–0.5	$(1.70 \times 10^{16})^a$
CO_2	Carbon dioxide	3.15×10^{-2}	2.45×10^{15}
Ne	Neon	1.82×10^{-3}	6.48×10^{13}
He	Helium	5.24×10^{-4}	3.71×10^{12}
CH_4	Methane	(1.5×10^{-4})	(4.3×10^{12})
H_2	Hydrogen	(5×10^{-5})	(2×10^{11})
N_2O	Nitrous oxide	(3×10^{-5})	(2×10^{12})
CO	Carbon monoxide	(1×10^{-5})	(6×10^{11})
NH_3	Ammonia	(1×10^{-6})	(3×10^{10})
NO_2	Nitrogen dioxide	(1×10^{-7})	(8×10^{9})
SO_2	Sulfur dioxide	(2×10^{-8})	(2×10^{9})
H_2S	Hydrogen sulfide	(2×10^{-8})	(1×10^{9})
O_3	Ozone	Variable	(3×10^{12})

Source: Walker, 1977.
Note: Total atmospheric mass equal to 5.14×10^{18} kg.
[a] Numbers in parenthesis are estimates.

concentration will double that of the preindustrial level by the middle of the next century. Business as usual refers to our continued reliance on burning fossil fuels and clear-cutting of temperate and tropical forests. Increased levels of atmospheric CO_2 and other greenhouse gases are expected to increase the global mean temperature by 1.5–5°C sometime between the years 2025 and 2050.

In addition to atmospheric gases, particulate matter is also present in the atmosphere and is made up of various organic or inorganic materials consisting of liquids or solids. Biological materials such as viruses, bacteria, spores, and pollen grains can also be classified as particulate matter if suspended in the atmosphere. Particulate matter is generally smaller than 0.5 mm in size and can be derived from several natural and anthropogenic sources. Particles that are less than 1 μm are capable of being retained in the atmosphere and transported long distances. Particles in the range of 0.001–10 μm are common in and around pollution source areas such as cities, highways, industrial centers, and power plants. Winds can greatly reduce the concentration of these materials within short distances from their source.

Airborne particles can originate from explosions, breakdown of materials by grinding action, volcanic activity, and wind erosion. Surface mining relies on explosives to loosen the underlying rock (Figure 2-2) so that it can be moved or processed. Large amounts of particulate matter can result from wind erosion in arid and semiarid areas. As an example, wind blown soil particles from arid regions have been estimated to contribute 10^{15} g of particulate matter to the atmosphere each year. Of this, 20% is less than 1.0 μm which can potentially be transported over long distances.

Trace elements are added to the atmosphere by several processes. Fossil fuel and coal burning, smelting of iron and nonferrous metals, volcanic ash, and wind erosion are responsible for increasing atmospheric concentrations of such elements as Au, Br, Cd, Pb, Se, Sn, and Te by as much as four orders of magnitude above the normal

Figure 2-2   Airborne particulates released after the detonation of explosives during a surface coal mining operation. Explosives are used to loosen the overburden materials that must be removed in order to mine the coal deposits.

levels. Trace elements can be bioaccumulated by some plants and microorganisms, and this can possibly increase the potential harmful effects of the trace elements. Trace element effects are discussed further in Chapter 7.

2.2.2 Atmospheric Cycles

Several elements and compounds have atmospheric cycles that are part of their overall transfer among soil, hydrosphere, and biosphere ecosystems. Some of the more important elements include C, N, P, and S. Cycles of N, P, and S are discussed in Chapters 4, 5, and 6, respectively. Chapter 10 provides information on the CO_2 cycle and its role in the greenhouse effect. In this section we will describe the basics of important atmospheric cycles, and provide general information to help one understand the interaction among the soil, water, atmosphere, and biosphere components.

Oxygen plays an important role in elemental cycles (i.e., C, N, P, and S). Oxygen is also a key element in atmospheric, geochemical, and life processes. Figure 2-3 indicates some of the various chemical reactions and fluxes that are involved in the oxygen cycle. Atmospheric oxygen (1.18×10^{21} g) represents the largest O_2 pool, which at present is in steady state due to consumption and production processes. Oxygen is consumed by respiration and organic matter decomposition, and is produced by photosynthesis. It has been suggested that all the O_2 in the atmosphere has been cycled through photosynthetic organisms such as plants and certain microorganisms. A large pool of oxygen exists in the lithosphere as reduced forms and is slowly released by weathering reactions. Oxygen is also consumed during the burning

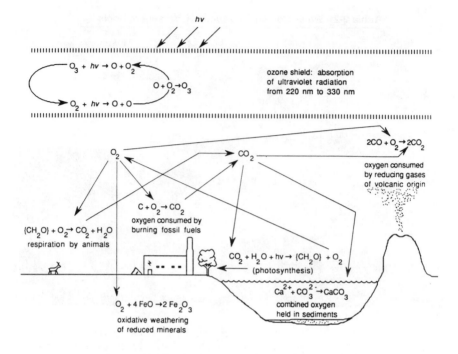

Figure 2-3   Examples of various oxygen and carbon dioxide reactions that occur in the environment. (Modified from Manahan, 1991.)

of fossil fuels, aerobic degradation, and oxidative weathering of soil, rocks and minerals.

Both O_2 and CO_2 cycles are regulated to a large extent by living organisms. The turnover rate or lifetime of O_2 and CO_2 in the atmosphere is related primarily to photosynthesis and respiration processes. The time required to cycle O_2 and CO_2 through the atmosphere is different due to the size of their pools. The *mean residence time* (MRT) is a measure of the time it takes a substance to cycle through a particular pool. The MRT of atmospheric O_2 as it cycles through the biosphere is approximately 3000 years; CO_2 takes only about 5 years.

Biogeochemical cycles of nutrient and trace elements can be extremely complex and are often studied by examining individual ecosystems (i.e., lithosphere, hydrosphere, atmosphere, and biosphere). Of primary interest are the transfer rates, MRTs, and fates of the elements in one or more of the various ecosystems. The influence of human activity on biogeochemical cycles has stimulated several studies to determine what the future may hold for us.

2.3 THE HYDROSPHERE

The hydrosphere includes water bodies such as rivers, streams, lakes, and oceans, as well as soil water, groundwaters, glaciers, and polar ice caps (Table 2-2). Although approximately 70% of the earth's surface is covered with water, an enormous amount

Table 2-2   Water Content in Various Hydrosphere Pools

Source	Content (km³)	Percent
Water bodies		
Oceans	1,320,000,000	97.2
Freshwater lakes	125,000	<0.1
Saline lakes and inland seas	104,000	<0.1
Rivers and streams	1,300	<0.1
Ice sources		
Polar ice caps and glaciers	29,200,000	2.2
Lithosphere		
Soil water	67,000	<0.1
Groundwater	8,350,000	0.6
Atmosphere	13,000	<0.1

Source: Goldman and Horne, 1983.

of water is hidden below ground. By far the largest source of water is the oceans, followed by ice (polar ice caps and glaciers) and then groundwaters. Freshwater and saline lakes and inland seas represent the largest pools of liquid water on land, with rivers and streams comprising only a small fraction of the world's water. The amount of water retained in soils is approximately 50 times that in rivers and streams. While the atmosphere contains only a small fraction of the water held in other pools, the amount of water that passes through the atmosphere is immense and extremely important.

The study of the chemical, physical, and biological properties and reactions of water bodies is called *hydrology; limnology* is the study of freshwater systems; *oceanography,* the study of the oceans; and *meteorology,* the study of climate and weather, which is highly dependent on water in the atmosphere. *Geohydrology* is the study of water in geological systems such as aquifers and groundwater environments. Soil scientists study the chemical, physical, and biological properties of soil ecosystems in which water plays a dominant role.

Water pollution implies that the quality of a water system has been degraded. Degradation refers to a decline in water quality by chemical, physical, or biological means, which is usually determined based on the intended use of the water. Water pollution can occur by a substance either directly or indirectly impacting the water system. Some of the more common pollutants include trace elements, nutrients, petroleum products, pathogens, and pesticides. Although suspended organic matter and sediments are generally present in most streams and lakes, these materials can cause water degradation by increasing the biological oxygen demand or decreasing light penetration by increasing water turbidity. Even changes in water temperature due to thermal discharges of industrial and power plants can alter biotic diversity in rivers and lakes.

Water quality is, and will continue to be, a major economic and environmental issue. The process of eutrophication is one example of why water quality is so important. Eutrophication of lakes and rivers occurs when excessive amounts of nutrients, such as N and P, are added to the ecosystem. As nutrient inputs to surface waters gradually increase, the trophic state of the water body passes through four trophic stages; *oligotrophic, mesotrophic, eutrophic,* and *hypereutrophic.* At each stage, progressive changes in the ecology of water bodies occur, usually affecting in

Table 2-3. Comparison of General Characteristics and
Select Properties of Oligotrophic and Eutrophic Water
Bodies.

Characteristic	Oligotrophic	Eutrophic
Nutrient status	Low	High
Algal blooms	Rare	Common
Biomass	Low	High
Aquatic diversity	High	Low
Dissolved oxygen (Saturation %)	>80	<10
Total P (µg/L)	4–10	>30
Total N (µg/L)	<200	>500
Chlorophyll (mg/L)	1–3	>8
Turbidity (m) (Secchi disk transparency)	6–12	<2

a negative manner their economic and recreational use. Excessive nutrient additions
to surface waters can lead to enhanced algal growth, decreased dissolved oxygen, and
reduced water transparency (Table 2-3). Some of the water quality problems associated with eutrophication are summarized in Table 2-4. See Chapters 4 and 5 for
further discussion on the roles of N and P in eutrophication.

2.3.1 Water Properties

Water is essential to all forms of life on earth. It is also the central component to
several soil chemical and physical processes. Ice can physically break down rocks
into small particles that then can be further weathered by chemical means. Dissolution of soil minerals, and their translocation, is a continual processes that is driven
by water leaching through the soil environment. In areas of low rainfall, weathering
and translocation of dissolved constituents is relatively slow when compared to high
rainfall areas. Transport of contaminants from soils to ground or surface waters is
generally hastened as the amount of water that percolates through the soil increases.

Water, which is often referred to as the universal solvent, is essential for the
transport of nutrients, gases, and organic compounds in living organisms. Some of
the unique properties of water are listed in Table 2-5. Many of these properties are
the result of the molecular structure of water and its ability to form hydrogen bonds.
Hydrogen bonding allows water molecules to interact with one another and form
clusters or liquid crystals. These interactions give water unique characteristics which

Table 2-4. Water Quality Problems Associated With Eutrophication.

Water quality problem	Contributing factors from eutrophication
Water safety, taste, odor	Nutrients, suspended sediments degrade water quality and increase cost and difficulty of drinking water purification; anoxic conditions and toxins produced in algal blooms can cause fish kills and make water unsafe for birds and livestock.
Low species diversity	Stimulated growth of certain organisms decreases number and size of population of other species; with time lake becomes dominated by algae and coarse, rapid-growing fish; high quality edible fish, submerged macrophytes and benthic organisms disappear
Impairment of recreational use and navigation	Increased sedimentation decreases lake depth, enhanced vegeta-tive growth blocks navigable waterways; decaying algal biomass produced surface scums, odors, and increases populations of in-sect pests.

Table 2-5   Properties of Water and Their Significance

Property	Significance	Comments
Solvent	Essential for many biochemical, chemical geologic, and atmospheric processes	Ubiquitous substance
Density	Allows ice to float on water	Maximum at density 4°C
Dielectric constant	Reasons why most ionic substances partially dissolve in water	Highest of all pure aqueous liquids
Surface tension	Produces unequal attraction forces between two phases	Highest of all aqueous liquids
Heat of evaporation	Controls rate of heat and water transfer between water and atmosphere	Highest of all substances
Latent heat of fusion	Stabilizes temperature change at freezing point	Highest of all aqueous liquids except NH_3
Heat capacity	Balances temperature changes	Highest of all aqueous liquids except NH_3
Transparency	Allows transfer of sunlight to great depths in water bodies where it is required by photosynthetic organisms	Colorless substance

Source: Manahan, 1991.

set it apart from other molecules of similar size and weight. For example, methane (molecular weight of 16) changes from a gas to a liquid at –55°C and from a liquid to a solid at –182°C.

2.3.2 Components of the Hydrologic Cycle

Water transfer or movement from one environment to another governs the hydrologic cycle (Figure 2-4). Water enters the atmosphere primarily by evaporation and transpiration processes and is returned to ocean and land surfaces in the form of rain and snow. The rate of water transferred from one pool to another, which is called

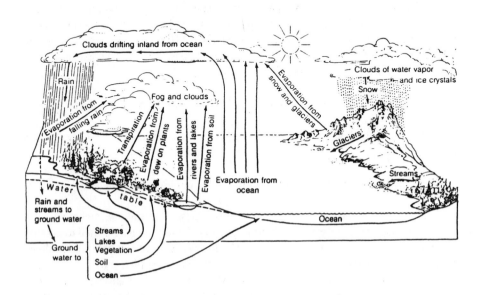

Figure 2-4   Illustration of the various pools and transfer processes that occur in the hydrologic cycle. (From Gilluly, et al., 1975. With permission.)

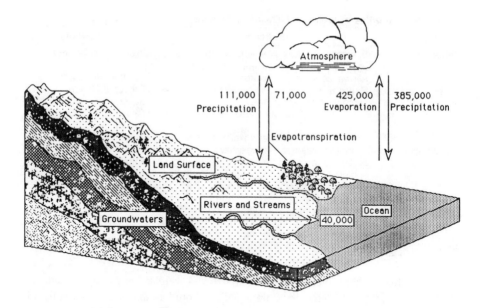

Figure 2-5 Global transfer rates (km³/year) for water movement in the hydrologic cycle. (Flux rates are from Spiedel and Agnew, 1982.)

water flux, is shown in Figure 2-5. According to Figure 2-5, the quantity of water leaving a water source is compensated by the amount of water entering. The rates shown are estimates for current water fluxes which have changed during the evolution of the earth.

The MRT for water in various pools can be calculated if it is assumed that input is equal to output and if the mass of the water in the pool and the rates at which water is entering and exiting the pool are known (i.e., MRT = mass/flux). Using the information in Table 2-2 and Figure 2-5 we can calculate MRT for the various pools. The MRT for water in the atmosphere would be:

$$\text{MRT} = 13,000 \text{ km}^3 \text{ / } 496,000 \text{ km}^3 \text{ yr}^{-1} = \quad 0.026 \text{ yrs or } 9.5 \text{ days} \qquad (2\text{-}1)$$

The fast rate at which water moves from land to oceans results in a MRT for streams and rivers of:

$$\text{MRT} = 1300 \text{ km}^3 \text{ / } 40,000 \text{ km}^3 \text{ yr}^{-1} = 0.033 \text{ yrs or } 12 \text{ days} \qquad (2\text{-}2)$$

Both of these examples indicate how sensitive these systems are, since a water-soluble pollutant entering either of these pools could be rapidly transported to another system over a short period of time. The *pollution potential* of a substance is related to its MRT in a particular system. Thus, if a contaminant spilled into a river due to an accident and the contaminant has a small MRT, it would be transferred rapidly to another pool. If the contaminant MRT in the receiving pool is larger, then the contaminant will remain in this pool for a longer period of time. As a comparison, the MRT of water in the ocean is:

$$\text{MRT} = 1,320,000,000 \text{ km}^3 / 485,000 \text{ km}^3 \text{ yr}^{-1} = 3,100 \text{ yrs} \qquad (2\text{-}3)$$

whereas, MRTs of water in lakes and groundwater systems have been estimated to be tens of years to hundreds or thousands of years, respectively.

2.3.2.1 Soils

Soils hold water in pore spaces by the cohesive and adhesive nature of water and soil particle surfaces. *Cohesion forces* are the result of water molecule polarity and hydrogen bonding, which attract water molecules to one another. *Adhesion forces* are responsible for attracting water molecules to soil mineral and organic matter surfaces. These forces allow water to move upward in soils by capillary action, or along surfaces of soil particles as water films.

Several terms are used to describe soil water, including water content and water potential. *Water content* is a measure of water in soils, usually expressed on a percentage basis. The water content of a soil is determined by determining the weight of both the water and solid materials contained in a soil sample. Each of these weights can easily be measured in the laboratory using an oven and a balance. The percent soil water content (SWC) is calculated as follows:

$$\% \text{ SWC} = [(\text{wet soil wt} - \text{oven-dried soil wt}) / (\text{oven-dried soil wt})] \times 100 \qquad (2\text{-}4)$$

Soil *water potential* is a measure of the strength, or energy, with which water is held by the soil. Water moves in soils from areas of high water potential to low water potential; water potential is in turn related to soil moisture content, textural class, structure, salt content, and organic matter content. Clays have lower water potentials than sands when both have the same water content. *Total water potential* is the overall effect due to a combination of several potentials, of which matric, pressure, gravity, and solute potentials are the most important. *Matric potential* (also known as tension or suction potential) is a combination of the interaction of water with soil surfaces and the tendency of small pores to retain water more strongly than large pores. In well-drained soils with low soluble salts, soil water potential is nearly equal to the matric potential. *Pressure and gravity water potentials* are related to external forces that are exerted on soil water. Pressure potential is due to atmospheric or gas pressure effects, and gravity potential is a result of the pull of gravity on soil water. *Solute, or osmotic, potential* is due to water movement from a dilute to a concentrated solution.

Water moves in soils as a vapor or a liquid. *Vapor flow* through a soil is generally a slow process. Water vapor is present in all unsaturated soils and moves by diffusion within the soil due to vapor pressure and temperature gradients. Soil water movement is classified as either saturated or unsaturated flow depending on the soil moisture content. Saturated flow occurs in soils where the void space is filled with water. Subsurface horizons can become saturated if water movement is restricted, for example, in soils with a high water table or a clay pan, or in stratified soils. Unsaturated flow occurs whenever void spaces are partially filled with air. In both saturated and unsaturated conditions, water flow is a function of the driving force acting on the water (*hydraulic gradient*) and the ability of the soil to allow water movement (*hydraulic conductivity*).

Water that infiltrates into soils can be stored; transferred to streams, rivers, lakes, oceans; or become part of the groundwater pool. Surface runoff occurs when rainfall cannot be absorbed by the soil because the rate of infiltration is slow or the soil becomes saturated. Water that falls on land surfaces can be returned to the atmosphere by evapotranspiration, which is a combination of evaporation from soil or plant surfaces and transpiration from plants. Runoff can result in soil erosion and pollution through the transport of soil particles which contain nutrients and possibly pesticides.

2.3.2.2 Inland Surface Waters

Inland surface waters include streams, rivers, and lakes. In general, water entering lakes comes from the surrounding area, which is known as its watershed or drainage basin. Sources of water entering streams and rivers can include: rainfall, surface runoff during periods of high rainfall, lateral water movement below the soil surface due to topography or stratified layers of different textures, water stored in adjacent wetlands areas, or groundwaters. Several natural (climate, vegetation, physiography, geology) and human (urbanization, agriculture, deforestation) factors influence the quality and quantity of water in inland surface waters.

There are several types of substances that can affect water quality. They include inorganic, organic, and biological materials; of these some have a direct impact on water quality, whereas others indirectly cause chemical, physical, or biological changes. Substances that can impact water quality, and which will be discussed later in this book, include N, P, S, trace elements, pesticides, acid rain, and greenhouse gases. Additional water pollutants such as radionuclides, carcinogens, pathogens, and petroleum wastes are also important in the context of environmental quality but will only receive limited coverage in this book.

When organic substances such as sewage sludge is added to surface waters, a rapid decline in available O_2 can occur. Oxygen is consumed in the biological decomposition of the added organic substances, and by oxidation of other reduced inorganic compounds (i.e., NH_4^+, Fe^{2+}, and SO_3^{2-}) present in the added material. This results in lower O_2 availability for higher forms of aquatic life. Two measures that are used to estimate the quality of surface waters are biochemical (or biological) oxygen demand (BOD) and chemical oxygen demand (COD). BOD is a measure of the amount of O_2 consumed by microorganisms over a 5-day period, whereas COD indicates how much oxidizable material there is in the sample by its reaction with dichromate. Values for COD are often higher than BOD, depending on the nature and quantity of oxidizable material in the water sample.

Biological communities (fish, plants, microorganisms, etc.) in surface waters are also impacted by conditions that are influenced by pH and salt concentrations. Mining activities can have a considerable effect on the quality of surface waters, as well as the land and air in the surrounding environment. Oxidation of reduced S substances can lead to acid mine drainage, which can be deleterious to plants, animals, and microorganisms. Acid mine drainage and some irrigation flow-through waters can also increase surface water salinity.

2.3.2.3 Groundwater

Infiltrating water can move below the soil environment into a region known as the vadose zone (unsaturated region) or move even deeper into the groundwater zone (saturated region) (Figure 2-6). The upper surface of the groundwater zone is called the water table, which fluctuates depending on the amount of water received by, or depleted from, the groundwater zone. The capillary fringe is the area above the water table where water in small pores is drawn upward by capillary action.

It was mentioned earlier that water movement is due to a combination of (1) hydraulic gradients and (2) ease with which water moves through soil or rock (hydraulic conductivity). Groundwater also responds to gradients, which are usually a product of gravitational force and permeability of substrata materials. Substrata are

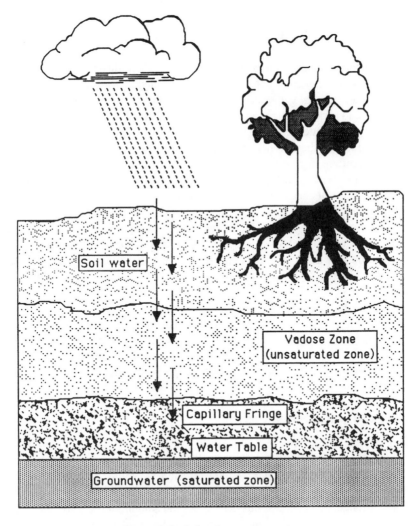

Figure 2-6 Subsurface water zones.

characterized by their porosity and permeability, which represent the degree of void space and resistance of water movement, respectively. Thus, groundwaters move faster in coarse-textured substrata and as the slope of the water table increases.

Aquifers are groundwater systems that have sufficient porosity and permeability to supply enough water for a specific purpose. In order for an aquifer to be useful, it must be able to store, transmit, and yield sufficient amounts of good quality water. Aquifers are classified as either confined (located under an impermeable substrata material) or unconfined (unrestricted above and having a water table). There may be enough pressure built up in a confined aquifer to create artesian conditions. Regions of substrata material that have low permeability and will not yield sufficient amounts of water to be practically useful are called *aquicludes* or *aquitards*.

2.4 THE SOIL ENVIRONMENT

The term soil has a different meaning to individuals in different scientific disciplines: to the agronomist or botanist, soil is best defined as a medium for the growth of plants; to the engineer, soil refers to the loose material that lies between the ground surface and solid rock; and to the soil scientist, soil is described as the unconsolidated mineral or organic matter at the earth's surface which has been altered by pedogenetic processes. Although there is no uniform definition for soil, it is apparent that the functions of soil are manyfold.

Soil pollution is often thought of as resulting from chemical contamination such as through the use of excessive amounts of pesticides and fertilizers that can result in surface water and/or groundwater contamination. However, there are other forms of soil pollution or degradation, including erosion, compaction, and salinity. Soils have often been neglected when they are used for on-site land disposal of waste chemicals and unwanted materials. Most soils are capable, to some degree, of adsorbing and neutralizing many pollutants to harmless levels through chemical and biochemical processes. There are limits, however, to soil being able to accept wastes without some negative effects.

Soils are a function of various physical, chemical and biological processes that are constantly at work changing soils over geologic time. In the following sections we will delineate the important features and define the terms commonly used to describe soils. In order to categorize soils for land-use purposes, one must understand the general properties of soils.

2.4.1 Soil Physical Properties

Soils contain solids, liquids, and gases. For soil physical properties, we are concerned primarily with what the composition and arrangement of solids are and how movement of liquids and gases are affected by solids. It should be noted, however, that soil color and temperature are also considered physical properties of soils. The arrangement of soil solids determines the amount of open volume, or *pore space,* that a soil possesses. In this section we will discuss the nature of soil solids and the importance of these solids in the movement of soil solutions and soil gases.

Table 2-6 Size Classification of Soil Particles According to
the U.S. Department of Agriculture System

Soil particles	Diameter (mm)	Comparison
Coarse fragments		
Stones	>254	>10 in.
Cobbles	75–254	3–10 in.
Gravel	2–75	0.08–3 in.
Soil particles		
Sand		
Very coarse	2.0–1.0	Thickness of a nickel
Coarse	1.0–0.5	Size of pencil lead
Medium	0.5–0.25	Salt crystal
Fine	0.25–0.10	Flat side of a book page
Very fine	0.10–0.05	Nearly invisible to the eye
Silt		
Coarse	0.05–0.02	Root hair
Medium	0.02–0.01	Nematode
Fine	0.01–0.002	Fungi
Clay		
Coarse	0.002–0.0002	Bacteria
Fine	<0.0002	Viruses

Soils are composed of solid materials ranging in size from stones to fine clays (Table 2-6). The larger materials, called coarse or mineral fragments (including stones, cobbles, and gravels), are chemically and physically weathered over long periods of time to form the smaller soil particles of sand, silt, and clay. Soil particles are defined on the basis of their diameter, although rarely do these particles exist as spherical objects. Clay minerals, for instance, are three-dimensional, layered structures that commonly have a platelike appearance. Soil particle sizes often differ with the classification schemes used by different groups; the U.S. Department of Agriculture system (Soil Survey Staff, 1975 and 1992) will be used throughout this book.

There are 12 soil texture classes that are defined by the relative proportion of sand, silt, and clay that comprise a soil sample (Figure 2-7). There are two generally used methods for determining soil texture: (1) the field method done by hand and (2) the mechanical analysis method. The field method is taught in introductory soil science courses and will not be discussed here. Using the mechanical analysis method to determine the sand, silt, and clay content requires the removal of coarse fragments by sieving the soil through a 2-mm sieve. Chemical treatments are also necessary to remove cementing agents such as organic matter and carbonates. The percentage of sand, silt, and clay should always total 100%; and once known, the soil texture can be found by using the textural triangle (Figure 2-7). For example, what is the textural class of a soil that contains 40% sand, 40% silt, and 20% clay? From the textural triangle, this soil would be classified as a loam. Only two of the soil particle percentages are actually needed to determine the soil textural class since the point at which the two meet does not change when the third particle percentage is used.

Because the soil textural class of a particular soil generally remains unchanged over time equivalent to the human life span, soil texture is often considered a basic property of the soil. This is one reason why soil descriptions used in soil surveys record the soil textural class of each horizon. However, a disturbance of an area such

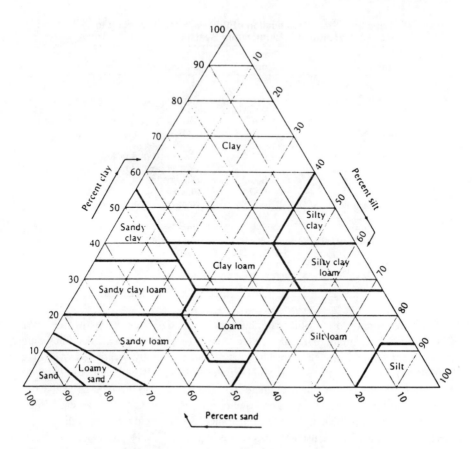

Figure 2-7   Textural triangle indicating the range in percent sand, silt, and clay for each soil textural class. The three corners of the triangle represent 100% of the primary soil particle-size classes. Note that a soil sample containing equal parts of sand, silt, and clay would be classified as a clay loam.

as water or wind erosion, could alter the textural class of the soil surface of both the soil being eroded and the soil receiving the erosional deposition. Over long time periods (geologic time), weathering and translocation of soil materials may change soil texture.

Soil particles that are held together by chemical and physical forces form stable *aggregates*. Natural aggregates are called *peds*. Collectively, the type of soil aggregates or peds defines *soil structure*. Soil structure influences the amount of water that enters a soil *(infiltration)* and gas diffusion at the soil surface. Soil structure also plays an important role in the movement of liquid and gaseous substances through soils. *Porosity* is a function of soil structure.

Soil structure is classified based on type, size, and grade. Common structure types include granular, platy, subangular and angular blocky, prismatic, and columnar shapes (Figure 2-8). Granular structure is representative of A horizons. Platy structure, although not very common, is found in E horizons of forest soils or in some clay-compacted B horizons. The blocky, prismatic, and columnar structures are often

Granular

Aggregates are small spherical peds typically found in most surface mineral horizons.

Platy

Horizontal plate-like structure often found in E horizons are clay pan soils.

Angular blocky

Sharp block-like peds commonly found in B horizons of humid region soils

Subangular blocky

Block-like peds with rounded edges also commonly found in B horizons of humid region soils.

Prismatic

Column-like peds with flat tops commonly found in B horizons of arid and semiarid soils.

Columnar

Column-like peds with rounded tops also commonly found in B horizons of arid and semiarid soils.

Figure 2-8 Soil structural types, descriptions, and their location in the soil profile.

found in B horizons; angular and subangular blocky structures are typical of soils in humid regions; and prismatic and columnar structures are common to soils in arid and semiarid regions. Structural size classes vary with the type of structure considered, and range from very fine to very coarse. Structural grade is determined by observing the soil structure in place in a soil pit and by determining how well-developed structure is. Structural grades include weak, moderate, and strong classifications.

Particle and bulk density measurements are useful for estimating the type of soil minerals present and the degree of soil compaction, respectively. *Particle density* is the mass of a particle per volume (Mg/m^3 or g/cm^3). Pore space and the weight of water are not included in particle density measurements. Common soil minerals (quartz, feldspars, micas, and clay minerals) have particle densities between 2.60 and 2.75 Mg/m^3. A value often used to represent the average soil particle density is 2.65 Mg/m^3. *Bulk density* is a measure of mass per volume of a soil. Undisturbed soils are usually used for bulk density measurements so that a true representation of amount of solid present in a particular soil volume can be calculated. With disturbed soils, natural soil pore spaces are destroyed. Bulk density is calculated on an oven-dried weight basis and does not take into consideration the amount of water present in the

Table 2-7 Common Primary and Secondary Minerals Found in Soils

Mineral	Chemical formula	Weatherability
Primary Minerals		
Quartz	SiO_2	Most resistant
Muscovite	$KAl_3Si_3O_{10}(OH)_2$	↑
Microline	$KAlSi_3O_8$	
Orthoclase	$KAlSi_3O_8$	
Biotite	$KAl(Mg,Fe)_3Si_3O_{10}(OH)_2$	
Albite	$NaAlSi_3O_8$	
Hornblende	$Ca_2Al_2Mg_2Fe_3Si_6O_{22}(OH)_2$	
Augite	$Ca_2(Al,Fe)_4(Mg,Fe)_4Si_6O_{24}$	↓
Anorthite	$CaAl_2Si_2O_8$	
Olivine	$(Mg,Fe)_2, SiO_4$	Least resistant
Secondary minerals		
Geothite	$FeOOH$	Most resistant
Hematite	Fe_2O_3	↑
Gibbsite	$Al(OH)_3$	
Clay minerals	Aluminosilicates	
Dolomite	$CaMg(CO_3)_2$	
Calcite	$CaCO_3$	↓
Gypsum	$CaSO_4 \cdot H_2O$	Least resistant

Source: Brady, 1990.

soil at the time of sampling. Mineral soils, unless developed in volcanic ash, generally have bulk densities greater than 1.0 Mg/m^3 with a common bulk density of a loam soil being 1.3 Mg/m^3. Soils with high densities (e.g., 2.0 Mg/m^3) will likely have slow water infiltration and permeability, which can result in ponding or surface runoff. Organic soils typically have bulk densities less than 1.0 Mg/m^3. Mineral soils derived from volcanic ash often have bulk densities in the range of 0.3–0.85 Mg/m^3.

Soil minerals are classified as primary and secondary minerals due to their origin. *Primary minerals* are those that formed during the cooling of molten rock and are predominately silicate minerals (Table 2-7). Igneous rocks are composed entirely of primary minerals; metamorphic and sedimentary rocks can contain abundant amounts of either primary and secondary minerals. *Secondary minerals* are formed in the soil from soluble products derived from the weathering of rocks. Clay minerals are one of the most important soil secondary minerals due to their large surface area and reactivity with ionic and dissolved organic compounds, and will be discussed in more detail in the next section. Carbonates and sulfates accumulate in B horizons and are the dominant secondary minerals present in soils in arid and semiarid regions.

Carbon is present in all soils in the form of organic matter. Soil organic matter contents vary from less than 1% in course-textured soils and soils of arid regions to nearly 100% in some poorly drained organic soils; typical farmland topsoils may contain 2–10% organic matter. Organic matter influences soil physical, chemical, and biological properties. Soil structure is often improved with the addition of organic materials such as manures, sludges, composts, and crop residues that are returned to the soil.

2.4.2 Soil Chemical Properties

Mineral solubility, soil reactions (pH), cation and anion exchange, buffering effects, and nutrient availability are major chemical properties of soils. These are determined primarily by the nature and quantity of the clay minerals and organic matter present. A knowledge of the chemistry of soil solutions is important to understanding how to handle waste materials such as manures, sewage sludges, and food by-products. In this section we will examine clay mineral and organic matter properties, and describe the soil reactions that are controlled by these materials.

2.4.2.1 Clay Minerals

Layered aluminosilicate minerals, better known as clay minerals, have a profound influence on many soil chemical reactions because of their high "active" surface area (Figure 2-9). The term active refers to charges that develop on clay mineral surfaces and the ability of some types of clay minerals to expand. Clay minerals should not be confused with clay-size particles, since these latter materials can also include particles of quartz, calcite, and gypsum that are 2 μm or less.

The clay minerals of most interest are crystalline and have regular layers of tetrahedral and octahedral sheets (Figure 2-10). Tetrahedral sheets are comprised of silicon and oxygen atoms with three out of every four oxygen atoms shared between adjacent tetrahedra. These shared oxygens are referred to as basal oxygens of the tetrahedral sheet. The remaining unshared oxygen is called the apical oxygen. Octahedral sheets are of two types — dioctahedral and trioctahedral. Dioctahedral sheets have two out of every three octahedral sites occupied, most often by the trivalent Al cation. Trioctahedral sheets have all octahedral sites occupied with divalent cations, which are commonly Mg.

The layered silicate clay minerals have structures that are either 1:1, 2:1, or 2:1:1 layers of tetrahedral and octahedral sheets. The 1:1 clay minerals have one tetrahedral and one octahedral sheet held together by the sharing of the apical tetrahedral oxygen

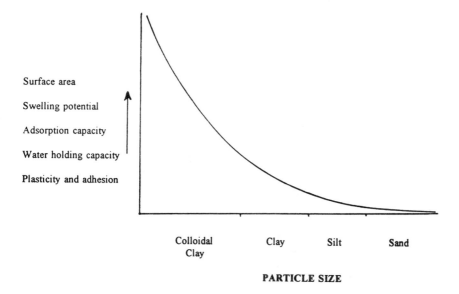

Figure 2-9   Relationship between particle size (using an equivalent weight or volume of material) and several soil chemical and physical properties.

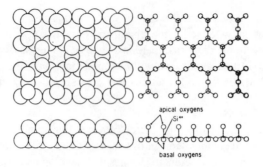

Tetrahedral sheet

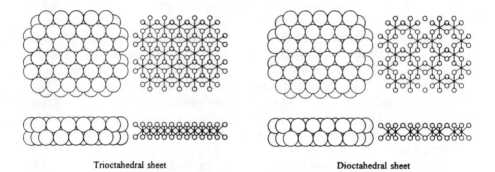

Trioctahedral sheet   Dioctahedral sheet

Figure 2-10 Two representations (space-filled and bond arrangement) of tetrahedral and octahedral sheets. Octahedral sheets can be trioctahedral or dioctahedral depending
on the number of sites filled. (Adapted from Schulze, 1989. With permission.)

(Figure 2-11). The 2:1 clay minerals have an octahedral sheet sandwiched between
two tetrahedral sheets (Figure 2-11). The 2:1:1 layered clay minerals are similar to
2:1 clays with an additional dioctahedral or trioctahedral sheet between the 2:1
layers.

The principal clay minerals found in soils are listed in Table 2-8. Kaolinite and
halloysite are 1:1 clay minerals that have Si in their tetrahedral sites and Al in
octahedral sites. The layers of the 1:1 clay minerals are held rather tightly together
through hydrogen bonds that form between basal oxygen of the tetrahedral sheets and
hydroxyls of the adjacent octahedral sheet. Kaolinite is a common mineral in soils
whereas halloysite is less stable and is often transformed to kaolinite over time.

Isomorphic substitution occurs when an element substitutes for another in the
mineral structure, such as Al^{3+} substituting for Si^{4+}. If an element of a lower charge
substituted for an element of a higher charge, a permanent negative charge develops
in the clay mineral. Both kaolinite (Figure 2-12) and halloysite have very little
surface charge due to low isomorphic substitution in their structures. Illite, or
hydrous mica, is a 2:1 nonexpanding, dioctahedral clay mineral which has K in the
interlayer spaces (i.e., area between the 2:1 layers) (Figure 2-12). The K atom is held
tightly in the interlayers and is commonly referred to as being *nonexchangeable*.
Montmorillonite is one of several types of clay minerals in the *smectite* family

PLANES OF IONS   SHEETS, LAYERS

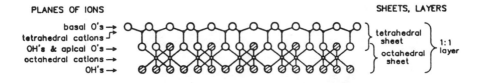

1:1 layer - one tetrahedral sheet and one octahedral sheet

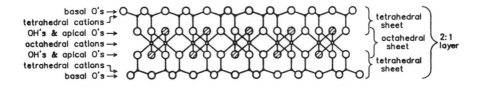

2:1 layer - 2 tetrahedral sheets with an octohedral sheet in the middle

Figure 2-11 Structural representation of 1:1 and 2:1 layers showing the chemical bonds between elements. (Adapted from Schulze, 1989. With permission.)

(Figure 2-12). Smectites are 2:1 clay minerals that have low to moderate isomorphic substitution. Because of the low permanent charge, smectites are capable of expanding, thus increasing the amount of exposed surface area. The high surface area together with the amount of charged sites gives smectites the ability to adsorb cations. *Cation exchange capacity* (CEC) is the amount of exchange sites which can adsorb and release cations. Montmorillonite is a common smectite in most soils. Vermiculite (Figure 2-12) is a 2:1:1 clay mineral which has high isomorphic substitution. The interlayers of vermiculite contain exchangeable Mg ions. However, since vermiculite has a higher layer charge, interlayer Mg in vermiculite is held more tightly than is interlayer Mg in smectites. Although vermiculite has a high CEC, when it is saturated extensively with K^+ or NH_4^+ ions, it becomes nonexpanding. Chlorite (Figure 2-12) is prevalent in many soils. The interlayer space between the 2:1 layers in chlorite is occupied by Mg octahedral sheets, thus making this a 2:1:1

Table 2-8 Common Clay Minerals Found in Soils Throughout the World[a]

Mineral	General formula
1:1	
Kaolinite	$Al_4Si_4O_{10}(OH)_8$
Halloysite	$Al_4Si_4O_{10}(OH)_8 4H_2O$
2:1	
Illite	$Al_4(Si_6Al_2)O_{20}(OH)_4 nH_2O$
Smectites	
Montmorillonite	$M_x(Al_{4-x}Mg_x)Si_8O_{20}(OH)_4$
Beidellite	$M_xAl_4(Si_{8-x}Al_x)O_{20}(OH)_4$
Nontronite	$M_xFe_4(Si_{8-x}Al_x)O_{20}(OH)_4$
Vermiculite	$Mg(Al,Fe,Mg)_4(Si_6Al_2)O_{20}(OH)_4 nH_2O$
2:1:1	
Chlorite	$Mg_6(OH)_{12}(Mg_5Al)(Si_6Al_2)O_{20}(OH_4)$

[a] M represents a monovalent cation.

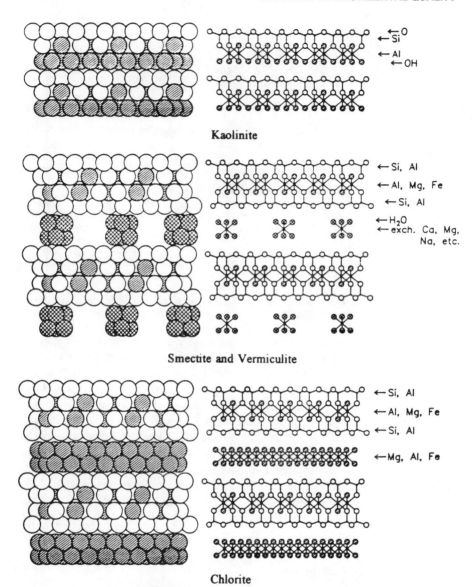

Figure 2-12 Structural representation of kaolinite (1:1), smectite and vermiculite (2:1), and chlorite (2:1:1) clay minerals. (Adapted from Schulze, 1989. With permission.)

layered silicate. Chlorites have low surface area due to the attraction of the brucite layer (Mg octahedral sheet) by the 2:1 layers, which reduces their potential CEC.

Figure 2-13 lists the general conditions that influence the formation of clay minerals and oxides. Some primary aluminosilicate minerals are physically and chemically weathered to produce clay minerals. Muscovite altered to illite is a good example of this. Both muscovite and illite have 2:1 structures; however, physical breakdown, loss of K (–K), and slight alteration in the muscovite structure produce

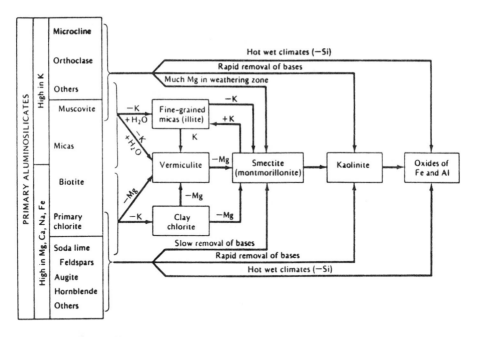

Figure 2-13 Sequence for the formation of clay minerals and Fe and Al oxides. Note the loss of soluble cations in the genesis of clay minerals. (From Brady, 1990. With permission.)

illite which has a lower CEC and greater swelling potential. Some clay minerals are recrystallized products of ions released during the breakdown of other minerals. Formation of kaolinite, a 1:1 clay mineral, is the result of recrystallization since there are no primary or secondary minerals with analogous 1:1 structures. As noted in Figure 2-13, kaolinite forms directly from the ions released from the weathering of primary aluminosilicates or smectite if accompanied by a rapid loss of cations. Intense weathering is responsible for the loss of Si and the production of Fe and Al oxides.

2.4.2.2 Organic Matter

Organic matter plays an important role in the chemistry of soils. Soil properties associated with soil organic matter include soil structure, macro- and micronutrient supply, cation exchange capacity, and pH buffering. Additionally, organic matter is also a source of carbon and energy for microorganisms.

Soil organic matter is comprised of decomposed plant and animal residues. It is a highly complex mixture of carbon compounds that also contain N, S, and P. Organic matter is made up of humic substances and biochemical compounds. Humic substances are operationally defined based on their solubility characteristics: humic acids are soluble in bases, but not acids; fulvic acids are soluble in acids and bases; and humin is the insoluble material that remains after humic and fulvic acid extraction. Biochemical compounds include identifiable organic compounds such as organic acids, proteins, polysaccharides, sugars, and lipids. General properties of soil organic matter and their effect on soils are listed in Table 2-9.

Table 2-9 Soil Organic Matter Properties and Their Associated Effect on Soil

Property	Remarks	Effect of soil
Color	The typical dark color of many soils is caused by organic matter	May facilitate warming
Water retention	Organic matter can hold up to 20 times its weight in water	Helps prevent drying and shrinking. May significantly improve the moisture-retaining properties of sandy soils
Combination with clay minerals	Cements soil particles into structural units called aggregates	Permits exchange of gases Stabilizes structure Increases permeability
Chelation	Forms stable complexes with Cu^{2+}, Mn^{2+}, Zn^{2+}, and other polyvalent cations	May enhance the availability of micronutrients to high plants
Solubility in water	Insolubility of organic matter is because of its association with clay. Also, salts of divalent and trivalent cations with organic matter are insoluble. Isolated organic matter is partly soluble in water	Little organic matter is lost in leaching
Buffer action	Organic matter exhibits buffering in slightly acid, neutral, and alkaline ranges	Helps to maintain a uniform reaction in the soil
Cation exchange	Total acidities of isolated fractions of humus range from 300 to 1400 cmol/kg	May increase the cation exchange capacity (CEC) of the soil. From 20 to 70% of the CEC of many soils (e.g., Mollisols) is caused by organic matter
Mineralization	Decomposition of organic matter yields CO_2, NH_4^+, NO_3^-, PO_3^{4-}, and SO_2^{4-}	A source of nutrient elements for plant growth
Combines with organic molecules	Affects bioactivity, persistence and biodegradability of pesticides	Modifies application rate of pesticides for effective control

Source: Stevenson, 1982. With permission.

From an environmental quality standpoint, organic matter can be both beneficial and detrimental. Soil organic matter can adsorb trace element pollutants (e.g., Pb, Cd, Cu) which will reduce the chance of contamination of surface- and groundwaters. Another advantage is the adsorption of pesticides and other organic chemicals. This reduces the possibility of pesticide carryover effects, prevents contamination of the environment, and enhances both biological and nonbiological degradation of certain pesticides and organic chemicals. In addition, organic matter is known for its capacity to adsorb inorganic (e.g., NO and NO_2) and organic gases (e.g., CO).

Although there are many benefits of soil organic matter, there are also detrimental effects that occur under certain situations. Soils with high organic matter contents may require higher pesticide application rates for effective control. Water contamination is then a concern if these pesticides are leached or transported by wind or water erosion.

Soils also have a finite capacity to adsorb trace elements and should not be used for long-term applications of organic wastes such as sewage sludge and manures that contain high levels of trace elements. Repeated applications of wastes containing trace elements (see Chapter 7) can lead to levels that may be toxic to plants, and possibly to animals and humans consuming of foods grown in the contaminated soils.

2.4.2.3 Ion-Exchange

Ion exchange is one of the most significant functions of soils. When isomorphic substitution occurs, cations of lower charge substitute for cations of high charge, and a permanent charge develops. Charged sites can also develop as a function of pH. Surface charges can form on clay minerals, metal oxides, and organic matter. These charged sites are the result of ionization (H^+ dissociation) or protonation of uncharged sites; ionization results in a negative-charged site, and protonation results in a positive-charged site. Both of these reactions are dependent on pH and are called pH-dependent charges. As the pH increases, the CEC of the soil is generally greater due to an increase in the amount of pH-dependent charged sites. Under acid soil conditions, some clay minerals, metal oxides, and organic matter will have positively charged, anion exchange sites. Inorganic and organic ions having charges that are opposite of the exchange site are attracted to the soil surface.

Both clay minerals and organic matter have the ability to sorb soluble chemicals from the soil solution. In addition, colloidal-sized clay minerals and dissolved organic matter can enhance the leaching and transport of heavy metals, organic chemicals (e.g., pesticides), and other potential pollutants.

2.4.3 The Biosphere

The living portion of soil consists of plants, animals, and microorganisms. Soil communities vary from soil to soil and include plant roots, rodents, worms and insects which are usually visible to the eye, and microorganisms (bacteria, actinomycetes, fungi, algae, and protozoa) which are so small they require a microscope to see them. Factors such as rainfall, temperature, vegetation, and physical and chemical properties of soils influence the type and number of living organisms in a soil community. The relative number and total biomass (weight of organism per unit volume) are given in Table 2-10 for a surface soil (note the extremely large number of microorganisms).

Organisms inhabiting soil environments can be grouped into two broad categories. These categories include the *autotrophs,* which assimilate C from CO_2 and obtain energy from sunlight or through the oxidation of inorganic compounds; and the *heterotrophs,* which use organic C as a source of energy and C. The autotrophs are considered *producers* because of their ability to convert CO_2 and energy from the sun into organic C products, a process called *photosynthesis.* Only vascular plants and some bacteria and algae are considered producers. Heterotrophs, on the other hand, are regarded as *consumers* and *decomposers.* Soil animals and most microorganisms fall into this category.

2.4.3.1 Plants

Aboveground plant parts are a source of food for consumers, such as grazing animals and humans. Below ground plant parts (i.e., roots, tubers, and other organs) are a source of food for humans and animals as well as soil consumers and decomposers,

Table 2-10 Estimated Number and Biomass of Soil
Animals and Microorganisms in Surface Horizons

	Abundance		Biomass
Organisms	(per meter^3)	(per gram)	(kg/HFS)
Soil animals			
Earthworms	200–2000	<1	110–1100
Nematodes	10^7–10^8	10^4–10^5	11–110
Others	10^4–10^6	Variable	17–170
Microorganisms			
Bacteria	10^{14}–10^{15}	10^8–10^9	450–4500
Actinomycetes	10^{13}–10^{14}	10^7–10^8	450–4500
Fungi	10^{11}–10^{12}	10^5–10^6	1120–11200
Algae	10^{10}–10^{11}	10^4–10^5	56–560
Protozoa	10^{10}–10^{11}	10^4–10^5	17–170

Source: Brady, 1990.
Note: Biomass values based on live weight per hectare furrow
slice (HFS).

and influence the type and activity of microorganisms living in and around the plant
roots. The *rhizosphere* is the area around roots that is influenced by the presence of
the roots; it often has from 10 to 100 times more microorganisms than the bulk soil.
Root exudates include organic and inorganic substances that provide nutrients for the
rhizosphere microorganisms.

Symbiotic associations between plant roots and microorganisms result in a variety
of beneficial effects to both the plant and the microbes. Nodules that form on
leguminous plant roots are caused by bacteria that are capable of converting atmospheric N_2 into N compounds utilizable by the plant. Mycorrhizal fungi form symbiotic relationships with plant roots. Plants benefit from this association by increased
absorption of nutrients (e.g., N, P, S, and micronutrients) and water.

2.4.3.2 Soil Animals

Soil animals include all animals that in one way or another influence soil properties. Large animals such as cattle and deer influence soils primarily through overgrazing. Effects of overgrazing include reduced ground cover, altered plant species
composition, soil compaction, and possibly soil erosion. Soil animals of a smaller
size that have a great impact on soil properties are those that burrow into soils.
Gophers, shrews, prairie dogs, badgers, moles, ground squirrels, and mice are some
of the burrowing animals that can profoundly influence the development of soils by
mixing topsoil and subsoil material. Tunnels and holes produced by burrowing
animals will increase the infiltration of water and air. Subterranean soil animals
include arthropods (e.g., ants, beetles, centipedes, millipedes, mites, spiders, spring-
tails), and worms (e.g., nematodes, earthworms), and protozoa. Nematodes are
generally the most abundant of all soil animals (Table 2-10 and Figure 2-14).

Several of the arthropods are involved in the decomposition of plant materials and
also assist in mixing the fragmented parts into the soil. Millipedes are primarily
saprophytic and consume dead organic materials. Ants and termites, as well as prairie
dogs, can mix large volumes of soil through their tunneling efforts.

Earthworms play an important role in soil formation and breakdown of organic
residues. They are usually found in the organic-rich surface horizon, but some

species burrow as deep as 6 m. Earthworms ingest organic matter and soil, which are decomposed by digestive enzymes and grinding action as they pass through the worms. Each day, the weight of soil material ingested and excreted by earthworms can be equivalent to their weight. The amount of earthworm casts (excreted material) produced yearly in a hectare of land has been estimated to range from 70 to 250 Mg/ha. Earthworm casts are rich in available nutrients such as N, P, K, Ca, and Mg when compared to the original soil material. *Lumbricus terrestris,* the common earthworm, is one of over 1800 worm species known worldwide.

Nematodes are microscopic, unsegmented roundworms that are classified based on their feeding habits. The most common are those that feed on decaying organic

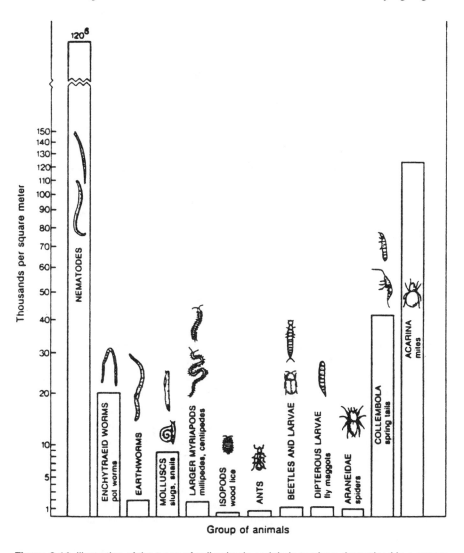

Figure 2-14 Illustration of the types of soil animals and their numbers determined in a square meter of a grassland soil. (From Paul and Clark, 1989. With permission.)

matter; whereas others prey on bacteria, fungi, algae, or other nematodes or they infect plant roots. One method of controlling the latter type of nematodes is by soil fumigation; however, this method of control can be very costly and potentially harmful to the environment.

2.4.3.3 Soil Microorganisms

Soil microorganisms include bacteria, actinomycetes, fungi, algae, and protozoa. *Viruses,* which are restricted to growing only in other living cells, are molecules of ribonucleic acid (RNA) or deoxyribonucleic acid (DNA) within protein coats and are discussed here only because they play an important part in the microbiology of soils. Viruses require a viable metabolizing host organism in order to reproduce or duplicate. They induce the host organism to manufacture the necessary components for virus reproduction. Many of the viruses can cause diseases such as those found in plants (e.g., tobacco mosaic and potato leaf roll) and animals (e.g., foot and mouth disease and bovine leukemia). Survival of the viruses that are pathogenic to humans and animals, and which are known to be in sludge wastes and manures, is a concern when these materials are land applied.

Bacteria are the most numerous of all soil microorganisms (Table 2-10), and individual bacteria cells are the smallest and most difficult to see under the microscope. A handful of soil can contain several billion bacterial cells. The ability of bacteria to rapidly reproduce and adapt to new environmental situations is important to the decomposition and transformation of both natural and anthropogenic (manmade) products. Some of the functions performed, either entirely or in part, by bacteria include: nutrient cycling, decomposition of organic materials, N fixation, pesticide detoxification, and oxidation-reduction reactions. Without bacteria to mediate these processes, life as we know it would not be possible.

Bacteria are classified as either autotrophs or heterotrophs, depending on their sources of energy and C. They can be further grouped as aerobes (require O_2), *anaerobes* (do not require O_2), and *facultative anaerobes* (grow in the presence or absence of O_2). The type and abundance of any one group of bacteria depends on such factors as available nutrients and soil environmental conditions. Some of the aerobic and anaerobic bacteria form spores that allow them to survive under adverse environmental conditions during dry and high temperature periods.

Soil conditions that affect the growth of bacteria are many; however, the most important factors are the oxygen, temperature, moisture, acidity, and inorganic and organic nutrient status of the soil environment. Both bacteria and fungi are usually active in aerated soils; but in ecosystems containing little or no O_2, most of the biological and chemical transformations are governed by bacteria. Optimum temperature and moisture requirements for most bacteria are 20–40°C and 50–75% of the moisture-holding capacity of the soil, although specialized groups are active at higher and lower temperatures and moisture contents. Extremely high and low pH (i.e., alkaline and acid conditions) are generally not suitable for most bacteria, but there are specialized bacteria that are active in soils with pH levels less than 3 and others above 10.

Actinomycetes physiologically resemble bacteria; however, their slender branched filaments also resemble filamentous fungi. In most soils, bacteria and fungi are more numerous than actinomycetes (Table 2-10); but in some warm climate soils, actino-mycete biomass may exceed that of others. While actinomycetes may not be as important as bacteria and fungi for carrying out biochemical processes, they are known to be involved in the decomposition of some resistant components of plant and animal tissues (e.g., cellulose, chitin, and phospholipids); synthesis of humiclike substances from the conversion of plant remains and leaf litter; breakdown of organic wastes (i.e., green manures, compost piles, and animal manures), even at high temperatures; infection of plants, animals, and humans; and excretion of antibiotics or production of enzymes that can influence soil community composition. The "earthy" aroma of soils is largely a volatile product of actinomycete activity.

Fungi obtain their energy and C from the decomposition of plant, animal, or soil organic matter. Two groups of fungi common to soils are the unicellular organisms called yeasts and the multicellular filamentous organisms (such as molds, mildews, smuts, and rusts), which are the most prevalent. Filamentous fungi are abundant in well-aerated fertile soils, whereas yeasts inhabit anaerobic environments. The fila-mentous bodies, which individually are called *hyphae* and collectively *mycelia,* are found interwoven among soil particles, organic matter, and plant roots. Mushrooms are the true reproductive structures of a filamentous fungi that may have extensive hyphae below ground.

Fungi perform several functions in soils including decomposing plant and animal organic substances, binding of soil particles into aggregates, forming symbiotic (mycorrhizal) associations with plants, and acting as predators in controlling certain microorganisms and soil animals. In acid surface layers and forest soils, fungi make up the majority of the soil biomass and are most active in the decomposition process.

Algae, like vascular plants and some bacteria, are capable of performing pho-tosynthesis because they contain chlorophyll. In soil with moist and fertile condi-tions, algae can potentially produce several hundred kilograms of organic sub-stances per hectare annually. Blue-green algae are capable of fixing atmospheric N_2 into organic N compounds. Lichens, which are a symbiotic association of an algae and a fungus, are commonly the first colonizers of bare rock and soil parent material. Organic acids synthesized and released by lichens are known to weather rock surfaces. They play an important role in the early stages of soil development in some areas.

Protozoa are unicellular animals that are primarily microscopic in size. Some protozoa, however, are known to reach macroscopic dimensions. They feed on decomposing organic matter and organisms such as bacteria and sometimes other protozoa. Protozoa populations are often related to bacteria numbers; as bacteria increase following the addition of organic residues to soils, protozoa increase as well.

2.4.4 Soil Development

Soil is a natural, three-dimensional array of vertically differentiated material at the surface of the earth's crust. The variation in soils throughout the world is a function of five soil-forming factors (Jenny, 1941), which can be expressed as:

$$Soil = f(pm, r, cl, o, t) \qquad (2\text{-}5)$$

with pm, r, cl, o, and t representing parent material, relief (topography), climate, organisms (vegetation), and time, respectively. Equation 2-5 indicates that the formation of a particular soil is determined by the amount of time a parent material located on a specific landscape has been affected by climate and organisms. *Soil genesis* is the process in which soil develops from parent materials and includes the physical and chemical weathering of parent material particles, physical movement of the particles, mineral alterations and transformations, addition of organic matter, and formation of horizons. Soils vary in horizon type and thickness, texture, structure, color, and other characteristics as shown in Figure 2-15.

Parent materials are mineral and organic materials that are chemically, physically, and biologically weathered to form soils. They are classified as either residual or transported materials. Residual parent materials are formed from the weathering of confined rocks and consist of igneous, sedimentary, and metamorphic rocks that vary in hardness, color, mineralogy, particle size, and crystallinity. Transported parent materials have been carried by wind or water; examples of these materials include eolian deposits of windblown sand, silt, and clay. Rapid moving water or slow moving glaciers are both capable of transporting large amounts of materials. Water-transported materials are generally sorted by particle size while glacial deposits, known as till, are often a mixture of particle sizes.

Figure 2-15 Examples of soil profiles representing different soil orders: (a) inceptisol (typic xerochept, loamy-skeletal, mixed, thermic) from Orange County, California; (b) spodosol (typic haplorthod, sandy, mixed, frigid) from Kalkaska County, Michigan; (c) mollisol (arenic argiustoll, fine, montmorillonitic, mesic) from Crook County, Wyoming.

Relief, also called topography, refers to the angle and length of the slope and proximity of the surface to the water table. The amount of water that infiltrates or runs off a soil is often determined by position of the soil in the landscape. The greater the angle and the longer the slope, the higher the probability for erosion. In addition, the direction of the slope (slope aspect) determines the amount of solar radiation received, which influences soil and air temperatures. In midlatitudes where the effects of slope aspect are greatest, the slope aspect often determines the type of vegetation growing on south- and north-facing sides of mountains. In the Rocky Mountains, forests can often be observed on north-facing mountain slopes due to cooler temperatures and higher soil moisture levels, whereas rangeland containing sagebrush can typically dominate south-facing slopes.

In discussing *climate* effects on soil formation, we are primarily concerned with how precipitation and temperature influence weathering and degradation rates, and translocation processes. The form (rain and snow) and amount of precipitation determine how much water may enter a soil; however, the topography of an area can cause water to concentrate on portions of the landscape. Also, water may erode the soil surface in which case soil degradation occurs. Weathering and degradation rates are generally faster at higher temperatures; thus a combination of increased precipitation and warmer temperatures will enhance soil formation and organic matter decay. Snowmelt can also increase the leaching of dissolved inorganic and organic chemicals that have built up over the course of the winter.

Plants, animals, and microorganisms are considered part of the *organism* soil-forming factor. Plants cycle nutrients taken from the atmosphere and soil and convert them to plant tissue, which after dying is decomposed by soil microorganisms and is recycled back to the soil. Organic matter is derived from both plant decomposition and synthesis reactions by soil microorganisms and plays an important role in weathering, degradation, and translocation processes. Some soil organic compounds can influence pH and the movement of dissolved ions to lower depths. Respiration of CO_2 by plant roots and microorganisms also has a profound effect on soil pH, especially in areas where intense respiration occurs. Physical mixing of soil particles and organic matter by soil animals can either enhance or decrease soil formation. Ants are capable of bringing course-textured material up from below the soil surface to build their mounds. Worms are excellent soil-forming animals known for their ability to enhance soil fertility and physical properties.

Soils are often described based on their morphological characteristics. Soils contain layers approximately parallel to the land surface; these are called *horizons* and are distinguished from one another due to differences in texture, color, structure, or other chemical and/or physical property that set them apart from the other horizons (Figure 2-15). Horizons can be formed from the loss *(eluviation)* of material such as organic matter, clay, or iron and aluminum; or from the accumulation *(illuviation)* of these constituents. Horizons are described first by a *master horizon* designation using the capital letters O, A, E, B, C, or R; and second, depending on the soil horizon characteristics, by a *subordinate designation* (Table 2-11). Surface horizons are classified as either O or A horizons depending on their organic matter content. In highly eroded areas where the soil surface materials have been removed, E, B, C, or R horizons may be exposed. Soils with E horizons are very common in forest soils. The E horizon is usually located below the O or A horizon, and forms as a result of the loss of considerable amounts of organic matter, clay, or iron and aluminum compounds *(sesquioxides)*. An E horizon is generally lighter in color than the horizons above and below. Horizons which have accumulated constituents translocated from horizons above are called B horizons. Parent material, the geologic material from which the soil has formed, is designated as a C horizon. Consolidated rock which is unaltered is classified as an R horizon. In addition to the five master horizons described above, examination of some soils indicate there are some horizons that appear to have characteristics similar to the horizons above and below. These are called *transitional horizons* and are designated by two capital letters such as AB or E/B.

Horizons are further subdivided to differentiate the major properties or characteristics that set them apart from other horizons. The use of the subordinated designation allows us to distinguish, for example, a B horizon that has accumulated clay material (Bt) from a B horizon that is only weakly developed (Bw). Not all subordinate designations are used with each of the master horizons. When describing the O horizon, one should always determine the decomposition state of the organic matter in order to correctly classify the horizon as an Oa, Oe, or Oi. These three subordinate

Table 2-11   Master Soil Horizons and Subordinate Designations

	Properties or characteristics
Master horizons	
O	Surface horizon dominated by organic matter
A	Mineral horizon with organic matter accumulation at the surface or just below the O
E	Horizon leached of organic matter, clay, and sesquioxides
B	Horizon with an accumulation of organic matter, clay, sesquioxides, carbonates, gypsum, and/or silica formed below the O, A, or E
C	Unconsolidated material underlaying A or B relatively unaffected by pedogenesis
R	Consolidated rock having little or no evidence of weathering
Subordinate designations	
a	Highly decomposed organic matter
b	Buried soil horizon
c	Concretions or nodules
d	High-density unconsolidated material
e	Partially decomposed organic matter
f	Frozen soil (permafrost)
g	Gleying (high water table)
h	Illuvial accumulation of organic matter
i	Relatively undecomposed organic matter
k	Accumulation of carbonates
m	Cementation
n	Accumulation of sodium
o	Accumulation of iron and aluminum oxides
p	Plowed or disturbed
q	Accumulation of silica
r	Weathered or soft bedrock
s	Accumulation of sesquioxides
t	Accumulation of silicate clays
v	Plinthite (red-colored iron material)
w	Weakly developed (color or structure)
x	Fragipan (high density)
y	Accumulation of gypsum
z	Accumulation of salts

designations are reserved solely for organic horizons. The subordinate designation p is only used with the master horizon letter A. Most of the subordinate designations are used to further classify B horizons. It is relatively easy to determine which of the lower case letters are used for B and sometimes C horizons since they represent processes of accumulation or changes from the geologic parent material of the soil.

2.4.4.1 Soil Classification and Land Use

There are several soil classification systems used throughout the world. The one used in the United States and in several other countries is *Soil Taxonomy,* which was developed by the U.S. Department of Agriculture (Soil Survey Staff, 1975). Soil Taxonomy was designed to group soils according to morphological characteristics and environmental properties such as diagnostic horizons, soil texture, soil structure, soil color, soil mineralogy, soil moisture and soil temperature regimes. It is a hierarchical system which contains six categories of classification ranging from a general grouping of all soils to a more specific, highly detailed grouping. The six categories, listed according to increasing detail, are: *order, suborder, great group, subgroup, family, and series.*

The initial step in classifying soils is to determine the *diagnostic horizons.* Diagnostic horizons are of two groups: *epipedons* — those forming at the soil surface — and *subsurface diagnostic horizons* — those forming below the soil surface (see Table 2-12 for horizon designations and properties). Of the six epipedon horizons listed in Table 2-12, only one — the histic epipedon which is designated as an O horizon — is composed primarily of organic matter. Of the 17 subsurface diagnostic horizons, the argillic, albic, cambic, and spodic horizons are common to soils in humid regions of the U.S. and Canada. Soils in arid parts of the U.S. often have one of the following diagnostic subsurface horizons: calcic, gypsic, natric, or petrocalcic. Knowledge of diagnostic horizons is essential for classifying soils.

Soils having similar types of diagnostic horizons are classified into 1 of 11 soil orders (Table 2-13). Knowing the soil order provides a general picture of what morphological and possible chemical and physical properties a soil possesses. The more detailed categories in Soil Taxonomy provide more specific information about the soil. The family name of a soil is the most useful for direct interpretation of soil properties and environment. Soil series are subdivisions of families, and are distinguished based on specific profile characteristics.

Soil survey reports (Figure 2-16) are a valuable source of information for land-use planning. They contain information that can help make land-use decisions for agricultural and nonagricultural purposes. In addition to delineating soil by map units, soil surveys contain detailed information for land-use decision making regarding the management of soils, croplands, woodlands, recreation areas, and wetlands. Soil surveys are also useful for evaluating soils for their suitability for irrigation and their drainage potential. Limitations for various land uses are listed in interpretive tables. These tables, along with the soil description, are a necessity for initial planning of waste disposal sites such as for septic tanks, sewage lagoons, and sanitary landfills.

Table 2-12   Diagnostic Horizons, Designations, and Their Properties

Diagnostic horizons	Designation	Properties
Epipedons		
Mollic	A	Thick, dark-colored, well-structured with high base saturation >50%
Umbric	A	Similar to mollic but with low base saturation <50%
Ochric	A	Light-colored, low organic matter that does not meet criteria of other epipedons
Anthropic	A	Man-made molliclike horizon high in phosphorus
Plaggen	A	Man-made thick horizon (>50 cm) developed from long-term manure applications
Histic	O	Organic horizon formed in poorly drained areas
Subsurface horizons		
Argillic	Bt	Accumulation of silicate clays as evidenced by clay films
Agric	A or B	Accumulation of organic matter, clay, or silt below the plow layer
Albic	E	Light-colored exuvial horizon which has lost organic matter clays
Calcic	Bk	Accumulation of calcite ($CaCO_3$) or dolomite [$CaMg(CO_3)_2$]
Cambic	Bw	Weakly developed horizon
Duripan	Bm	Hard pan due to cementing with silica
Fragipan	Bx	Weakly cemented, brittle layer
Gypsic	By	Accumulation of gypsum
Natric	Btn	Argillic horizon high in sodium having columnar or prismatic structure
Oxic	Bo	Highly weathered with accumulation of Fe and Al oxides and nonexpanding silicate clays
Petrocalcic	Bk	Cemented calcic horizon
Petrogypsic	By	Cemented gypsic horizon
Placic	B	Thin cemented plan that is black to red and held together by iron, manganese, and/or organic matter
Salic	Bz	Accumulation of salts
Sombic	B	Accumulation of organic matter low in base saturation
Spodic	Bh, Bhs, Bs	Accumulation of organic matter and sesquioxides
Sulfuric	B	Highly acid soil with sulfur-containing mottles

Source: Soil Survey Staff, 1975.
Note: Diagnostic horizons are used to classify soils at high levels of Soil Taxonomy.

Table 2-13   Soil Orders and Their Diagnostic Features According to
Soil Taxonomy

Soil order	Common diagnostic horizon	Diagnostic features
Alfisol	Argillic/natric	Base saturation >50%; no mollic, oxic, or spodic
Andisol	None required	Volcanic ash derived soil; noncrystalline or poorly crystalline minerals
Aridisol	Ochric	Dry soil; argillic, cambic, or natric common
Entisol	Ochric	Nominal profile development
Histosol	Histic	Organic soil; peat or bog
Inceptisol	Cambic	Few diagnostic features; ochric or umbric horizons common
Mollisol	Mollic	Dark soil with high base saturation; no oxic or spodic
Oxisol	Oxic	Highly weathered; no argillic or spodic
Spodosol	Spodic	Accumulation of organic matter and sesquioxides; albic horizon common
Ultisol	Argillic	Base saturation <50%; no oxic or spodic
Vertisol	None required	Contains high amounts of swelling clay causing deep cracks when dry

Figure 2-16 Example of a map sheet taken from the Soil Survey of Goshen County, Wyoming. The mapping units listed on the map represent the type of soil association, which is a landscape that has a distinctive pattern of soils, and slope of the area. For example, the city of Torrington is located primarily on map unit HnA which represents Haverson and McCook loams on 0–3% slopes. (From U.S. Department of Agriculture, Soil Conservation Service in Cooperation with the Wyoming Agricultural Experiment Station, 1971.)

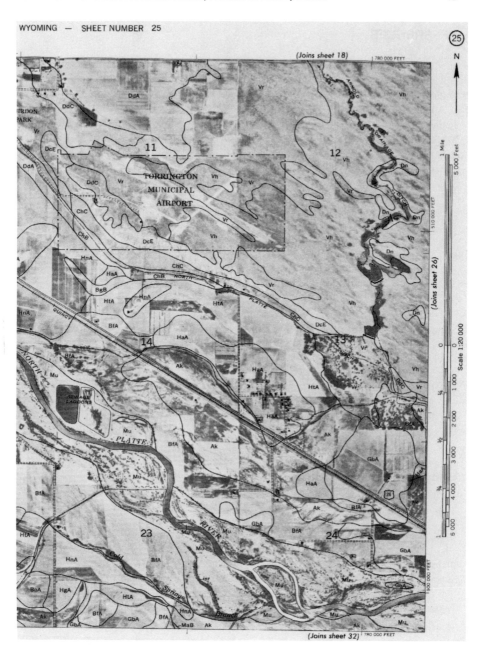

Figure 2-16 (Continued)

REFERENCES

Brady, N. C., *The Nature and Properties of Soils,* 10th ed., Macmillan, New York, 1990.

Goldman, C. R. and Horne, A. J., *Limnology,* McGraw-Hill, New York, 1983.

Gilluly, J., Waters, A. C., and Woodford, A. O., *Principles of Geology,* 4th ed., W. H. Freeman, San Francisco, CA, 1975.

Jenny, H., Factors of Soil Formation, McGraw-Hill, New York, 1941.

Manahan, S. E., *Environmental Chemistry,* 5th ed., Lewis Publishers, Chelsea, MI, 1991.

Paul, E. A. and Clark, F. E., *Soil Microbiology and Biochemistry,* Academic Press, San Diego, CA, 1989.

Schlesinger, W. H., *Biogeochemistry: An Analysis of Global Change,* Academic Press, San Diego, CA, 1991.

Schulze, D. G., 1989. An introduction to soil mineralogy, in *Minerals in the Soil Environment,* 2nd ed., Dixon J. B. and Weed, S. B. Eds., Soil Science Society of America, Madison, WI, 1989, 1.

Soil Survey Staff, Soil taxonomy: a basic system of soil classification for making and interpreting soil surveys, *Agricultural Handbook No. 436,* U.S. Government Printing Office, Washington D.C., 1975.

Soil Survey Staff, *Keys to Soil Taxonomy,* 5th ed., Technical Monograph No. 19, Soil Management Support Services, Pocahontas Press, Blacksburg, VA, 1992.

Spiedel, D. H. and Agnew, A. F., *The Natural Geochemistry of Our Environment,* Westview Press, Boulder, CO, 1982.

Stevenson, F. J., *Humus Chemistry: Genesis, Composition, Reactions,* John Wiley & Sons, New York, 1982.

U.S. Department of Agriculture, Soil Conservation Service (in cooperation with the Wyoming Agricultural Experiment Station), Soil Survey of Goshen County, Wyoming: Southern Part, by Stephens, F., Brunkow, E. F., Fox, C. J., and Ravenholt, H. B., 1971.

Walker, J. C. G., *Evolution of the Atmosphere,* Macmillian, New York, 1977.

SUPPLEMENTARY READING

The Atmosphere

Bach, W., Pankrath, J., and Kellogg, W., Eds., *Man's Impact on Climate,* Elsevier, New York, 1979.

Benarie, M. M., Ed., *Atmospheric Pollution,* Elsevier Publishing, New York, 1982.

Sloane, C. S. and Tesche, T. W., Eds., *Atmospheric Chemistry: Models and Predictions for Climate and Air Quality,* Lewis Publishers, Chelsea, MI, 1991.

Tegart, W. J. G., Sheldon, G. W., and Griffiths, D. C., *Climate Change: The IPCC Impacts Assessment,* Australian Government Publishing Service, Canberra, Australia, 1990.

The Hydrosphere

Berner, E. K. and Berner, R. A., *The Global Water Cycle,* Prentice-Hall, Englewood Cliffs, NJ, 1987.

Hites, R. A. and Eisenreich, S. J., *Sources and Fates of Aquatic Pollutants,* American Chemical Society, Washington, D.C., 1987.

Mance, G., *Pollution Threat of Heavy Metals in Aquatic Environments,* Elsevier Applied Science, Essex, England, 1987.

Van der Leeden, F., Troise, F. L., and Todd, D. K., *The Water Encyclopedia,* 2nd ed., Lewis Publishers, Chelsea, MI, 1990.

The Soil Environment

Dixon, J. B. and Weed, S. B., Eds., *Minerals in the Soil Environment,* 2nd ed., Soil Science Society of America, Madison, WI, 1989.
Keller, E. A., *Environmental Geology,* Merrill Publishing, Columbus, OH, 1988.
Page, A. L., Logan, T. J., and Ryan, J. A., Eds., *Land Application of Sludge: Food Chain Implications,* Lewis Publishers, Chelsea, MI, 1987.
Singer, M. J. and Munns, D. N., *Soils: An Introduction,* 2nd ed., Macmillian, New York, 1991.
Smith, M. A., Ed., *Contaminated Land: Reclamation and Treatment,* Plenum Press, New York, 1985.

3   CLASSIFICATION OF POLLUTANTS

3.1 DEFINING POLLUTION AND CONTAMINATION

Before a general discussion of the major environmental issues in soil science can be presented, a definition of the terms *pollution* or *pollutant* must be given. Interestingly, definitions of these terms will vary from individual to individual. Consulting a dictionary yields synonyms such as impure, unclean, dirty, harmful, or contaminated. Although these words are appropriate, they do not provide a working definition that is useful for the study of the environment.

Part of the problem in defining pollutant or pollution is that an agreement must be reached on what constitutes acceptable use of materials that may cause pollution. For example, some would consider the use of pesticides acceptable if they were reasonably certain that the effect of the pesticide is only what was intended, whereas others would not consider any use of pesticides acceptable. In the first case, the pesticide is a pollutant only if undesirable side effects occur; while in the second case, the pesticide is always a pollutant. What constitutes an acceptable level of pollution is another way to view this problem. Attitudes will range from none *(ecocentric)* to any level that will not harm me *(egocentric)*.

A second part of the problem in defining pollutant or pollution is the distinction between *anthropogenic sources* and *natural sources*. A volcano may place a greater quantity of noxious gases and particulate matter into the atmosphere than the combined output from a large number of electric power plants, but some would not consider the output from the volcano as a pollutant because of its natural origin. Similarly, heavy metal mining activities may pollute soils with heavy metals, yet soils with high concentrations of heavy metals can occur naturally because of their proximity to metal ore deposits.

It should be apparent that there are value judgments associated with defining pollutant or pollution, and it is impossible to change this situation. A reasonable working definition of a *pollutant,* taking into account the aforementioned problems, would be *a chemical or material out of place or present at higher than normal concentrations that has adverse affects on any organism.* The implications of this definition are that the pesticides applied to agricultural soils are not pollutants provided they do not move below the rooting zone of the crop or run off the surface. They would become pollutants if they occurred off-site at concentrations high enough to cause harm to an organism. The volcanic output and the naturally occurring, metal-contaminated soils represent pollutants because a chemical or material is present at higher than normal concentrations and can have adverse affects on organisms. Miller (1991) states, "Any undesirable change in the characteristics of the air, water, soil, and food that can adversely affect the health, survival, or activities of humans or other living organisms is called pollution," which also takes into account noise or thermal pollution in addition to pollution caused by chemicals or materials. Both definitions allow for value judgments by using the subjective phrases "out of place" and "undesirable change."

The term *contaminated* is often used synonymously with *polluted,* although subtle differences in the definitions would indicate that this practice is not correct. Contaminated implies that the concentration of a substance is higher than would occur naturally, but does not necessarily mean that the substance is causing harm of any type. In addition, polluted refers to a situation not only in which the concentration of a substance is higher than would occur naturally, but also in which the substance is causing harm of some type (as noted in the two definitions given previously). By this reasoning, a soil could be contaminated but not polluted.

The terms *toxic waste, hazardous waste,* or *hazardous substance* are often heard; and obviously such materials can be pollutants. The added distinctions of *toxic* or *hazardous* are used for substances that can be acutely or chronically toxic to humans, as opposed to pollutants such as sediment or phosphorus, which are not. There is no regulatory classification for toxic waste from a legal standpoint in most states. *Hazardous waste,* defined as part of the Resource Conservation and Recovery Act (RCRA), is solid waste that causes harm because of its quantity, concentration, or physical, chemical, or infectious characteristics. Harm is defined as significantly contributing to an increase in mortality, or serious illness, or presenting a significant hazard to humans or the environment if improperly treated, stored, transported, or disposed. The cover of the March 1985 issue of *National Geographic* magazine depicted hazardous waste as a dark, pasty substance being handled by an individual in full protective gear. This aptly summarized the image that most people have of hazardous waste. Most of us, however, have hazardous materials stored under our kitchen sinks under the guise of drain cleaners, scouring powders, glass cleaners, bleach, etc.

Pollution is often broadly categorized according to its source. *Point source pollution,* as the name implies, is pollution with a clearly identifiable point of discharge; for example, the outflow from a wastewater treatment plant or a smokestack. *Nonpoint source pollution* is pollution without an obvious, single point of discharge; for example, surface runoff of a commonly used lawn herbicide. The implications for control are quite different. Controlling nonpoint source pollution can be difficult because of the large areas involved and having to deal with multiple landowners and sources. One of the successes of the early versions of the Clean Water Act was a reduction in P discharges to surface waters from numerous point sources. These sources were primarily wastewater treatment plants. Federal matching money was made available and many communities upgraded their treatment plants from primary to secondary or advanced facilities. Sensitive water bodies, such as Lake Erie and Chesapeake Bay, benefited greatly from this. However, P in surface waters remains a significant problem today. It is primarily nonpoint in origin and

Table 3-1 Classification of Pollutants and the Impacted Medium

Pollutant category	Examples	Soil	Water Ground	Water Surface	Air
Nutrients	Nitrogen and phosphorus in commercial fertilizers, manures, sewage sludges, municipal solid waste	*	*	*	
Pesticides	Insecticides, herbicides, fungicides	*	*	*	
Hazardous substances	Fuels, solvents, volatile organic compounds	*	*	*	*
Acidification	Acid precipitation, acid mine drainage	*	*	*	*
Salinity and sodicity	Road salt, saline irrigation water	*	*	*	
Trace elements	Cationic metals, oxyanions, elements normally present at low concentrations in soils or plants	*	*	*	
Sediments	Soil lost via water erosion			*	
Particulates	Soot, soil lost via wind erosion, volcanic dusts				*
Greenhouse gases	Carbon dioxide, methane, nitrous oxide, chlorofluorocarbons				*
Smog-forming compounds	Ozone, secondary products of fuel combustion				*

comes from agricultural uses of commercial fertilizers and animal manures. This contribution to the P problem will be much more difficult to solve. This topic will be discussed in more detail in Chapter 5.

3.2 CLASSIFYING POLLUTANTS

Table 3-1 presents a broad classification scheme for pollutants based on their general characteristics or uses. Each pollutant category can impact more than one medium. Each category can also represent numerous processes or members. Trace elements, for example, encompass more than 20 different elements. Acidification consists of several unrelated reaction pathways, some occurring in the atmosphere and others in the lithosphere, that produce acidic products that negatively impact their respective endpoints. Conversely, sediments comprise a relatively straightforward category with a single member.

The *nutrients* category primarily reflects the negative impacts of nitrogen and P when present at high relative concentrations. Phosphorus in surface waters and nitrates in surface- and groundwaters are the main indicators of environmental problems. Agricultural production practices are responsible for a fair share of the problems associated with excessive nutrients. Significant contributions are also made from private home and horticultural uses of fertilizers, naturally occurring sources, sewage treatment facilities, septic tanks, food processing plants, and livestock.

The *pesticides* category represents a wide range of mostly organically based chemicals used to control pests such as weeds, insects, rodents, or plant pathogens (see Chapter 8). Once again, agricultural production practices account for the majority of pesticide use, with private home and horticultural uses also being significant. Pesticides are normally released intentionally into the environment at low levels, although accidental spills involving large quantities or high concentrations do occur.

Hazardous substances probably comprise the broadest category of all. It is a catchall term that basically includes materials that are acutely or chronically toxic to humans or other organisms when improperly administered, used, or disposed of and are not included in the remaining categories. Many of the hazardous substances are organic compounds with properties analogous to pesticides. However, hazardous substances are generally not intentionally released into the environment; and soil contamination problems usually involve improper use, disposal, or accidental spills over small areas with high concentrations of the chemical in question. Contrast this with the environmental concerns for the general use pesticides, which generally involve applications at low concentrations to large areas as an accepted practice.

The *acidification* category includes several unrelated processes. Acid precipitation is mainly the end result of conversion of the oxides of N and S into their respective acids in the atmosphere. The concern is with precipitation having a lower than normal pH and its impact on the environment. Acid mine drainage represents another source of acidification where water has been acidified by the weathering and oxidation of sulfide minerals, primarily pyrite (FeS_2). Such water can have a pH as low as 2.0 and can have significant impacts on soil or surface water systems. Another example is that of soils being acidified after heavy use of ammoniacal N fertilizers without neutralization of the acidity (from the nitrification process) with liming materials.

The primary problems associated with *salinity* and *sodicity* are reductions in plant productivity due to water stress caused by the increased osmotic potential and changes in soil physical properties (dispersion with a resultant reduction in permeability) when salt or sodium concentrations in soils become too high. Irrigation with water containing high salt concentrations or a high level of Na relative to Ca and Mg are the chief causes of these problems. Runoff from roads deiced with sodium chloride onto roadside soils is another lesser source of salinity and sodicity.

Trace elements are elements that are normally present in relatively low concentrations in soils or plants. They may be essential for growth and development of humans or other organisms, although many are not. Trace elements of concern as pollutants are those that cause acute or chronic health problems in humans, animals, plants, or aquatic organisms when present in above critical threshold concentrations (see Chapter 7).

Sediments represent soil particles that have eroded from the landscape and have been carried to surface waters. Areas most susceptible to erosion include construction sites, recently tilled farmland, or overgrazed pastures. Sediments can physically block light transmission through water, which can alter the ecology of the body of water; can be enriched in P, which can accelerate eutrophication; and can act as sources or sinks for a variety of water pollutants. Sediments also can accumulate in surface waters, thus inhibiting navigation and recreation activities and reducing the longevity of dam structures.

Particulates, greenhouse gases, and *smog-forming compounds* are air pollutants. Particulates are relatively inert particles generally consisting of carbon, soil, volcanic ash, etc. suspended in the atmosphere. Greenhouse gases are the gases responsible for the greenhouse effect. The *greenhouse effect* is an increase in the concentration of various gases in the atmosphere — primarily CO_2, CH_4, N_2O, and chlorofluorocarbons — that make it more difficult for radiated heat to escape the atmosphere, resulting in an increase in the mean global temperature above that existing without the influence of man. Some greenhouse gases are man-made and others are naturally occurring, although their concentrations are increasing in the atmosphere due to human activities. Certain aspects of food production may also increase the concentrations of some greenhouse gases. Smog-forming compounds are the ingredients for the complex process of smog production.

Nutrients, pesticides, hazardous substances, trace elements, acidification, and greenhouse gases will be discussed in more detail in subsequent chapters. The topics of sediments, particulates, salinity, and smog-forming compounds are beyond the scope of this book.

3.3 SOIL QUALITY

An underlying theme in this text is *soil quality*. This can be measured in many different ways, depending on factors such as the desired use of the soil materials themselves or of a particular landform. In the broadest sense, soil physical and chemical properties determine soil quality. *Physical properties* such as bulk density and texture influence aeration, permeability, infiltratability, water-holding capacity, or constructive properties can be quantified and related to quality. *Chemical properties* are the concentrations of organic and inorganic constituents that determine characteristics such as soil fertility, biological activity, degree of pollution, salinity, corrosiveness, or shrink-swell potential and are also quantifiable and related to soil quality. Indeed, the majority of this book discusses soils with excessive concentrations of nutrients, pesticides, hazardous substances, or trace elements.

On a more practical scale, we consider issues such as the ability of a soil to produce quality food, fiber, or feed; the construction properties and limitations; and the ability to maintain an ecosystem or desired land use. These are sometimes referred to as the *soil functions*. The ability of a soil to produce quality food, fiber, or feed refers to the contribution the soil makes to the capability of a site to have profitable production of crops that are free of harmful substances. The *site quality* would then include a desirable combination of soil characteristics (including low concentrations of pollutants), climate, and available water. Construction properties and limitations, topics beyond the scope of this text, refer primarily to chemical and mineralogical characteristics that influence soil properties such as shrink-swell capacity and corrosiveness which, in turn, determine suitability for basements or buried materials.

The ability to maintain an ecosystem or desired land use can be explained with an example. A soil that has developed under a tropical rain forest has the ability to maintain that ecosystem and, by our reasoning, is of high quality. If the forest is removed and replaced with crops, profitable crop production can be maintained only for a few years. This is because of the rapid decline in soil organic matter, with an

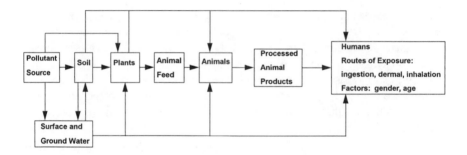

Figure 3-1   Pathways for human exposure to pollutants applied to soils. (Adapted from Brams et al., 1989.)

associated depletion of nutrients and reduction in soil physical qualities, due to exposure to oxidizing conditions and the rapid erosion that occurs as the high amounts of rainfall interact with the unprotected soil surface. After a short time, erosion and changes in soil fertility will likely have rendered that soil unable to maintain either the rain forest ecosystem or the desired land use (cropland); and its quality will be dramatically reduced. Contrast this with soils in the midwest region of the U.S., where hardwood forests were cleared for cropland. In this situation, the soil continues to be able to support profitable crop production and also could still support the native vegetation.

3.4 HUMAN EXPOSURE TO SOIL POLLUTANTS

Figure 3-1 presents a schematic of the pathways for human exposure to pollutants applied to soils. The major component of this flow chart is the human food chain, where crops are consumed directly by people or fed to livestock and the animal products are used as food. Surface- or groundwater sources of drinking water also are included. It is necessary to differentiate the various steps a pollutant goes through as it progresses from plant uptake to human consumption of processed animal tissue. As the pollutant is transferred from one compartment to another, various processes can influence the amount that eventually will reach the human endpoint. For example, regulations pertaining to maximum permissible concentrations of Pb and Cd in soils allow much higher concentrations of Pb as compared to Cd even though Pb is more toxic to humans than Cd. This is because, at equal concentrations of soil Pb and Cd, Pb concentrations in plants will be much lower than Cd and, consequently, food chain transfer of Pb will be less. Similarly, pollutants can be partitioned into different body parts within an animal. Fat, organ meats, and muscle tissue accumulate different concentrations of a particular pollutant. Polychlorinated biphenyls (PCBs), for example, are fat soluble and strongly partition into fatty tissue as compared to muscle tissue. Some investigators use transfer coefficients to describe the change in pollutant concentration as it is transferred from soil to plant, plant to animal feed, etc. Obviously, each pollutant would require a unique set of *transfer coefficients*, which can have values ranging from close to zero to as high as 1000 or even higher.

In addition to food chain transfer, several other pathways by which humans are exposed to soil pollutants are depicted in Figure 3-1. Pollutants can reach surface- or groundwaters and be reapplied to soil as irrigation water or be ingested directly by animals or humans. Grazing animals and infant children can consume soil directly. Humans inhaling soil dust particles are exposed to pollutants contained in the dust. Infants with hand-to-mouth activity or pica children (children who have an abnormal craving to eat substances not fit for consumption, especially soil or paint) can directly consume significant quantities of soil. These additional pathways must also be considered in regulatory decision making. Pica children, for example, could alter regulations for soil Pb levels because they are at greater risk than other individuals exposed via food chain transfer.

Figure 3-1 is a very simplified summary of some complicated processes. Conceptually, however, it covers the major steps involved in human exposure to soil pollutants. One could construct similar diagrams for any organism. References to Figure 3-1 will appear throughout this book.

3.5 MAJOR ENVIRONMENTAL ISSUES IN SOIL SCIENCE: A SUMMARY

One could assemble a very long list of individual environmental problems related to soil science. The list would include such problems as leaking underground storage tanks, trace element contaminated soils, saline soils, eroded soils, and acidified soils. It is probably more instructive to present categories of the major environmental issues in soil science. Most of the issues fall into one or more of the following categories:

1. **Reductions in soil quality because of unacceptable concentrations of pollutants** — This category includes soils that either directly or through food chain transfer expose humans or other organisms to pollutants that may cause direct detrimental effects. This category is quite extensive and includes problems associated with the pesticides, hazardous substances, and trace elements pollutant categories listed in Table 3-1.
2. **Reductions in soil quality that limit soil function** — Eroded, acidified, or salt-affected soils that can no longer support a desired land use or ecosystem fall into this category. Direct detrimental effects to humans or other organisms due to exposure to pollutants are generally not an issue in this case.
3. **Soils as a source of contaminants** — This category includes leaching and runoff losses of various chemicals or materials from the landscape. Here the presence of a substance in the soil is not the primary problem but rather the effects of the substance on the environment as it leaves the original point of application is of concern. Thus, a reduction in soil quality as indicated by chemical analysis is not necessarily the issue, but instead it is the conflict between soil function or use and the surrounding environment. Horticultural and agricultural uses of pesticides and nutrients would be prime examples here. Solving these problems ultimately requires an understanding of all risks and a prioritization of land uses and the desired quality of the environment.

Society's responses to environmental soil science issues are varied. Regulations are written that prevent soil contamination or that control the use of substances considered pollutants. Research is conducted on methods for remediating contaminated soils and on understanding the fate and transport of contaminants in the environment. Government programs are used to prevent degradation of soil quality by erosion. Considerable work needs to be done with regard to understanding the interaction of the soil environment with potential pollutants, and to utilizing the soil resource for the benefit of society while maintaining soil quality.

REFERENCES

Brams, E., Anthony, W., and Witherspoon, L., Biological monitoring of an agricultural food chain: soil cadmium and lead in ruminant tissues, *J. Environ. Qual.*, 18, 317, 1989.

Miller, G. T., *Environmental Science: Sustaining the Earth,* Wadsworth Publishers, Belmont, CA, 1991.

SUPPLEMENTARY READING

Hillel, D. J., *Out of the Earth, Civilization and the Life of the Soil,* The Free Press, A Division of Macmillan, New York, 1991.

Nebel, B. J., *Environmental Science: The Way the World Works,* Prentice-Hall, Englewood Cliffs, NJ, 1991.

Wagner, T. P., *Hazardous Waste: Identification and Classification Manual,* Van Nostrand Reinhold, New York, 1990.

4 SOIL NITROGEN AND ENVIRONMENTAL QUALITY

4.1 NITROGEN AND THE ENVIRONMENT

Nitrogen (N) is arguably the most important and yet most difficult to manage of all the plant nutrients. While absolutely vital to modern agriculture, it has also been shown to have a number of potentially serious environmental impacts, as briefly summarized in Table 4-1. This chapter will focus on N in agricultural soils, emphasizing the issues of greatest importance to environmentally sound soil N management. Some situations that differ considerably from traditional production agriculture (e.g., land reclamation, urban horticulture, forestry) will be reviewed to illustrate the variety of approaches needed to effectively manage soil N in an ecosystem shared by cities, farms, and industries. We will address the following key questions:

- What is the basis for public concerns about the effects of N on human and animal health, its role in the pollution of surface- and groundwaters, the formation of acid rain, and the destruction of the stratospheric ozone layer?
- How can we use our knowledge of the many complex chemical and biological processes of the *soil N cycle* to improve our management of all N sources, from fertilizers to animal manures, sewage sludges, and industrial organics?

4.1.1 Origin and Distribution of Nitrogen in the Environment

To fully understand the environmental problems caused by N, and to develop sensible, cost-effective approaches to N management, it is essential to have a basic understanding of the origin and cycling of N in the earth's four major "spheres": the lithosphere, hydrosphere, atmosphere, and biosphere (see Chapter 2 for discussion of these spheres). Most (>98%) of the earth's N is found in the lithosphere, either in the earth's core, in igneous and sedimentary rocks, in oceanic sediments, or in soils. The remaining 2% is distributed between the atmosphere, hydrosphere, and biosphere. In the atmosphere, N is found primarily as the inert gas N_2, which comprises 78% of atmospheric gases. In the hydrosphere, N occurs as dissolved organic or inorganic N. Nitrogen is also a vital component of the biosphere, which consists of living plants and animals. Nitrogen can be found in many different forms in these spheres, including molecular N, organic molecules, geologic materials, gases, and soluble ions.

Table 4-1   Summary of Environmental Problems Associated with Nitrogen

Environmental issue	Causative mechanisms and impacts
Human and animal health	
Methemoglobinemia	Consumption of high nitrate drinking waters and food; particularly important for infants because it disrupts oxygen transport system in blood
Cancer	Exposure to nitrosoamines formed from reaction of amines with nitrosating agents; skin cancer increased by greater exposure to ultraviolet radiation due to destruction of ozone layer
Nitrate poisoning	Livestock ingestion of high nitrate feed or waters
Ecosystem damage	
Groundwater contamination	Nitrate leaching from fertilizers, manures, sludges, wastewaters, septic systems; can impact both human and animal health, and trophic state of surface waters
Eutrophication of surface waters	Soluble or sediment-bound N from erosion, surface runoff, or groundwater discharge enters surface waters; direct discharge of N from municipal and industrial wastewater treatment plants into surface water; atmospheric deposition of ammonia and nitric acid; general degradation of water quality and biological diversity of freshwaters
Acid rain	Nitric acid originating from reaction of N oxides with moisture in atmosphere is returned to terrestrial ecosystem as acidic rainfall, snow, mists, or fogs (wet deposition) or as particulates (dry deposition); damages sensitive vegetation, acidifies surface waters, and — as with eutrophication — can unfavorably alter biodiversity in lakes, streams, bays
Stratospheric ozone depletion, global climate change	Nitrous oxides from burning of fossil fuels by industry and automobiles and from denitrification of nitrate in soils are transported to stratosphere where ozone destruction occurs; ultraviolet radiation incident on earth's surface increases as does global warming

Source: Keeney, 1982.

Nitrogen is a very dynamic element, capable of being transformed biochemically or chemically through a series of processes conceptually summarized as the *nitrogen cycle* (Figure 4-1). Most N transformations involve the *oxidation* (loss of electrons) or *reduction* (gain in electrons) by the N atom from both biological and chemical means. The oxidation states of N in nature range from +5 in the nitrate anion (NO_3^-) to −3 for ammonia (NH_3) or ammonium (NH_4^+). The *soil N cycle* (Figure 4-2), described in detail in Section 4.2, is a subset of the overall N cycle. In this chapter we seek to understand how management of the soil N cycle affects other segments of the global N cycle, such as groundwater aquifers in the hydrosphere.

In the broadest sense, there are three major natural inputs to the soil N cycle, *atmospheric deposition, biological N fixation,* and *weathering and decomposition.* Atmospheric deposition occurs as inorganic and organic N in precipitation or dry particulate matter. Biological N fixation is the conversion of gaseous, atmospheric N_2 to NH_3 and then organic N by symbiotic and nonsymbiotic organisms. Weathering and decomposition reactions are those in which previously fixed or deposited organic or inorganic N is transformed from stable inorganic or organic N to more chemically and biologically active forms. The major natural processes by which N is lost from the soil include evolution as a gas (NH_3 *volatilization, bacterial* or *chemical denitrification*

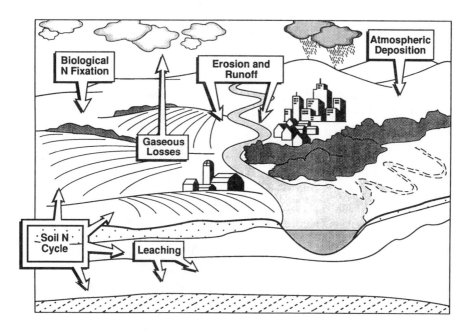

Figure 4-1 The global nitrogen cycle. Inset indicates soil N cycle, shown in more detail in Figure 4-2.

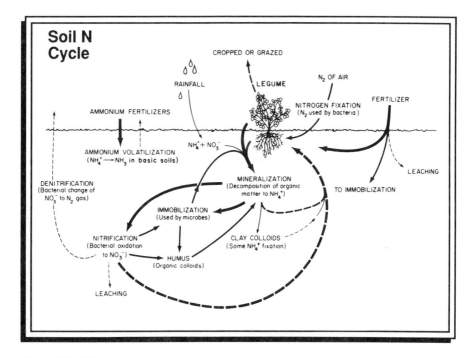

Figure 4-2 The soil nitrogen cycle. (Adapted from Donahue et al., 1983. With permission.)

of NO_3^-); and transport processes, as soluble N in waters flowing downward through soils *(leaching)* or in water and sediments moving across the soil surface *(erosion* and *runoff)*. Soil N management primarily involves manipulating or supplementing (e.g., fertilization) these natural processes to produce plants for food, fiber, or aesthetic purposes. Environmental concerns about N arise when one of these transformations results in the conversion and concentration of N in a form that can adversely affect the health or quality of an organism or an ecosystem. It is important to remember that, although only a few forms of N are now regarded as environmentally harmful, the processes of the N cycle regulate, on a global and local scale, the amount of N found in each of these forms. Controlling pollution caused by N, therefore, starts with an understanding of how we can control the N cycle in soils and other ecosystems.

Global estimates of long-term changes in the distribution of N among these four spheres are filled with uncertainty. However, human activities have clearly impacted this distribution, enriching some sectors of the earth's environment with N while simultaneously depleting others. Many fundamental aspects of modern civilization — such as agriculture, urbanization, industry, transportation, and water or waste treatment systems — have the potential to significantly affect the distribution of N on a localized scale and cumulatively on a global scale. Unfortunately, this human-induced movement of an element such as N to a part of the environment where it can have a negative effect is often a synonym for pollution. Although (as seen in Table 4-1) there can be a wide variety of environmental impacts from this redistribution of N, from the perspective of soil N management, the forms of N of greatest importance are nitrate (NO_3^-) and nitrous oxides (N_2O, NO). While our understanding of the mechanisms by which NO_3^- and nitrous oxides cause pollution are far from complete, we do have a basic understanding of the processes involved and the means by which this pollution can be avoided.

4.1.2 Nitrogen Effects on Human and Animal Health

There is little doubt that groundwater, and more specifically drinking water, contamination by NO_3^- is presently the environmental issue of greatest concern for N management. The major human and animal health issues associated with the consumption of excessive NO_3^- from drinking waters, or even certain foods that are part of human and animal diets, are *methemoglobinemia* ("blue baby syndrome") and possible carcinogenic effects from another class of nitrogenous compounds, the nitrosamines. Methemoglobinemia is not caused directly by NO_3^-, but occurs when NO_3^- is reduced to nitrite (NO_2^-) by bacteria found in the digestive tract of humans and animals. Nitrite can then oxidize the iron in the hemoglobin molecule from Fe^{2+} to Fe^{3+}, forming methemoglobin, which cannot perform the essential oxygen transport functions of hemoglobin. This can result in a bluish coloration of the skin in infants, hence the origin of the term blue baby syndrome. Methemoglobinemia is a much more serious problem for very young infants than adults because after the age of 3–6 months, the acidity in the human stomach increases to a level adequate to suppress the activity of the bacteria that transform NO_3^- to NO_2^-. Although documented cases of methemoglobinemia are extremely rare, the U.S. Environmental Protection Agency has established a maximum contaminant level of 10 mg N/L (45 mg NO_3^-/L) to protect the safety of U.S. drinking water supplies. Animals can also be susceptible to methemoglobinemia, although the health advisory level for most livestock is much higher, approximately 40 mg N/ L (180 mg NO_3^-/L).

The other major health concern with N is the potential carcinogenic effect of nitrosamines, compounds with the general chemical structure $R_2N–N=O$, where R represents a carbon group (e.g., CH_3^-). Nitrosamines are formed by the reaction, under highly acidic conditions (pH < 4.0), of secondary and tertiary amines (R_2NH, R_3N) with nitrous acid anhydride (N_2O_3). They have been shown to produce tumors in laboratory animals, but conclusive evidence for a causative role of nitrosamines in human cancer does not exist.

4.1.3 Nitrogen and Eutrophication

Eutrophication is defined as an increase in the nutrient status of natural waters that causes accelerated growth of algae or water plants, depletion of dissolved oxygen, increased turbidity, and general degradation of water quality. Causes and management of eutrophication are discussed in more detail in Chapters 2 and 5, but the enrichment of lakes, ponds, bays, and estuaries by N and P from surface runoff or groundwater discharge are known to be contributing factors. The levels of N required to induce eutrophication in fresh and estuarine waters are much lower than the values associated with drinking water contamination. Although estimates vary and depend considerably on the N:P ratio in the water, concentrations of 0.5–1.0 mg N/L are commonly used as threshold values for eutrophication. Marine environments, where salinity levels are greater, are more sensitive to eutrophication and thus have lower threshold levels of N (<0.6 mg N/L). It is important to remember that N in surface waters reflects not only agricultural inputs (primarily nonpoint in nature), but also inputs of N from direct discharge of wastewaters from municipalities, industry, and recreational developments. Atmospheric deposition, as both precipitation ("acid rain") and particulate matter, and fixation of atmospheric N by aquatic organisms also contribute to the total pool of N in surface waters.

4.1.4 Atmospheric Effects of Nitrogen: Acid Rain and the Ozone Layer

Nitrogen has been shown to have three serious effects on the earth's atmosphere, and because of these atmospheric changes, on the quality of terrestrial and aquatic environments. Nitric acid (HNO_3), primarily caused by the release of nitrous oxides (N_2O, NO) to the atmosphere during the burning of fossil fuels, is a major component of *acid rain*. Studies have shown that acid rain (pH < 5.6), acid mist, or dry deposition of acidic particulates can negatively and seriously affect forest ecosystems and surface waters, but the impact on agricultural soils and crops to date has been minimal.

Nitrous oxides have also been shown to cause the photooxidation of ozone (O_3) in the stratosphere, reducing the capacity of the ozone layer to protect the earth from the intense ultraviolet radiation emitted by the sun and contributing to global warming. A simplified version of the reactions of N oxides with ozone is shown below:

$$N_2O + O \xrightarrow{\text{Photodissociation}} 2NO \qquad\qquad (4\text{--}1)$$

$$NO + O_3 \xrightarrow{\text{Photooxidation}} NO_2 + O_2 \qquad\qquad (4\text{--}2)$$

In addition to the burning of coal, oil, and gasoline, a major nonpoint source of N_2O or NO is the process of biological *denitrification,* in which soil microorganisms reduce NO_3^- to N_2O under oxygen-limited conditions. *Chemodenitrification* can also occur in soils, resulting in the evolution of N oxides under well-aerated conditions. These processes are discussed in more detail in Section 4.2.2. The third effect is the role of N_2O in the anthropogenic greenhouse effect, which will be discussed in Chapter 10.

4.1.5 Risk Assessment for Nitrogen Pollution

The environmental effects of N described above have been documented on local, regional, and global scales. Unlike some forms of pollution, the adverse impacts of N are not directly obvious or dramatic and the level of risk is not as clear. Given the limited nature of the resources available to mitigate all forms of pollution and the critical importance of N to food production, proper assessment of this risk is essential to prioritize efforts to remediate N pollution. The process used for risk assessment will be described in detail in Chapter 11, but certain points are clear at this time with regard to N pollution and soil management.

First, there is a clear public perception that agriculture contributes to N pollution of surface- and groundwaters by improper fertilization and organic waste management. While scientific research has substantiated this concern, it has also indicated that groundwater contamination by NO_3^- is often localized in nature and associated with specific regional problems such as well-drained soils, shallow water tables, highly concentrated animal production, intensive irrigation, and waste or wastewater disposal by municipalities or industry. The combined weight of public perception and scientific documentation assures that the risk will be addressed by scientific, advisory, and regulatory agencies. In essence, there is sufficient agreement among all parties that risk exists and that a significant commitment of resources to reduce its impacts is needed. In contrast, atmospheric effects of N oxides originating from agricultural soils are much less understood. Further, the greater role of industry and urban areas in generating N oxides clearly dictates that, while the emission of N from soils should be minimized, our control efforts should first be directed toward nonagricultural sources.

Second, we must acknowledge that there are significant obstacles to reducing N pollution originating from both agricultural and nonagricultural soils. The pressures to produce increasing amounts of food with fixed amounts of arable land often result in the use of higher and often less efficient N rates as we substitute fertilizer N for soil N. The need to maintain farm profitability and the low costs of fertilizer N favor the use of insurance fertilization to overcome unexpected N losses that may be caused by uncontrollable climatic events. The nature of modern animal-based agriculture concentrates nutrients derived from soils, fertilizers, and organic N in other regions, into areas without an adequate land base for proper N use. A similar scenario exists for organic wastes produced in cities (e.g., sewage sludge). Conversion of grasslands and forests to agricultural land for the production of annual crops can release organic N stored for centuries over a relatively short period of time, enhancing the potential for enrichment of surface- and groundwaters with NO_3^-. Recreational developments near sensitive water bodies can discharge N from septic systems into groundwaters; overfertilization of turf in urban areas can produce runoff that is high in NO_3^-. It is apparent that the technology, education, effort, and cost required to design and implement improvements in N use for each of these scenarios is formidable. Prioritization of risk thus becomes of critical importance as we consider the allocation of resources.

Finally, it is also important to acknowledge the time required to correct the problem. Groundwaters that have been contaminated with NO_3^- by land-use practices over a 30-year period are likely to require several decades to "dilute" to acceptable concentrations with the low NO_3^- leachate coming from soils receiving improved management practices. This means that the cost of reducing N pollution of our natural waters must be borne by society for many years, and that a long-term, integrated effort between all responsible parties will be needed.

In summary, at this point in time we have acknowledged that N pollution is an important environmental issue, particularly for groundwaters, and one that requires complex, long-term solutions. This chapter will review the key components of the soil N cycle that must be managed to reduce N pollution.

4.2 THE SOIL NITROGEN CYCLE

The *soil nitrogen cycle* (Figure 4-2) is a subset of the global N cycle and can be viewed as a conceptual summary of interactions between the chemical, physical, and biological transformations undergone by N in soils. As this chapter focuses on environmental concerns related to soil N management, the interactions of the soil N cycle with those segments of the environment sensitive to pollution by N (groundwaters, surface waters, the atmosphere) will be emphasized. From this perspective, the key N transformations are the cycling of N between organic and inorganic forms *(mineralization and immobilization),* gaseous losses of N *(ammonia volatilization* and *denitrification)* to the atmosphere, losses associated with water movement *(leaching* and *erosion),* and *biological N fixation.* Many of these reactions are controlled by soil microorganisms that alter the form, oxidation states, and thus the fate of N between N_2, N_2O (nitrous oxide), NH_3/NH_4^+ (ammonia/ammonium), NO_2^- (nitrite), and NO_3^- (nitrate). It is critical to recognize, from a management standpoint, how the relative importance of these reactions varies with soil and environmental conditions and when it is possible to exert a significant degree of control over a given reaction. Our ability to control a N transformation, however, will depend not only on the biology, chemistry, or physics of the process, but also on the intensity and economics of management. Farmers using irrigation, for example, have greater control over the timing of N delivery to crops than those in dryland agriculture because of their ability to inject soluble N fertilizers into irrigation waters. This can improve the efficiency of N uptake by crops, reducing NO_3^- leaching. Industries involved in land reclamation or municipalities charged with sewage sludge disposal may increase the extent of waste processing (e.g., composting or lime stabilization) in response to greater regulation on organic waste use related to pathogens, heavy metals, or organic pollutants. These changes in waste properties may then significantly affect the N transformations likely to occur with land application of the waste.

Sound environmental management of N begins with a thorough understanding of the major components of the soil N cycle. Enormous scientific effort has been expended to study N transformations in soils. The challenge now is to translate this body of knowledge into practical management programs that achieve both production and environmental goals. In the following sections the basic principles underlying each of the key N transformations will be described, and in later sections of this chapter, techniques that can be used to manage these N transformations will be related to agricultural, urban, and land reclamation issues.

4.2.1 Mineralization, Nitrification, and Immobilization of N

Mineralization refers to the conversion of organic forms of N (e.g., proteins, chitins and amino sugars from microbial cell walls, and nucleic acids) to inorganic N, as ammonium-N (NH_4^+). The organic N may be indigenous to the soil or freshly added as crop residues, animal manures, or municipal wastes. The process is mediated by a diverse population of heterotrophic soil microorganisms (bacteria, fungi, actinomycetes; described in Chapter 2) that produce a wide variety of extracellular enzymes capable of degrading proteins (proteinases, peptidases) and nonproteins (chitanases, kinases) into NH_4^+. These microbes use the energy derived from the oxidation of soil organic matter for metabolic activities and the N released during the decomposition process to produce the amino acids and proteins essential for population growth. The reactions involved in mineralization of organic N to inorganic NH_4^+ can be summarized as follows:

$$\text{Organic N} \xrightarrow{\text{Proteolysis, Aminization}} \text{Amino} - N\left(R - NH_2\right) + CO_2 + \text{Energy, by-products} \quad (4\text{-}3)$$

$$\text{Amino} - N\left(R - NH_2\right) \xrightarrow{\text{Ammonification}} NH_3 + H_2O \rightarrow NH_4^+ + OH^- \quad (4\text{-}4)$$

Once mineralized, NH_4^+ can be taken up by plants, nitrified, immobilized by soil microorganisms, lost as a gas by ammonia volatilization, held as an exchangeable ion by clays or other soil colloids, or fixed in the interlayers of certain clay minerals. Each of these fates of NH_4^+ is described in the following sections.

Mineralization of N from soil organic matter has been shown to provide a significant portion of the N requirement of many crops. Plants can absorb NH_4^+ directly from the soil solution; and, in fact, many studies have shown that NH_4^+ is taken up preferentially by plants over other sources of N (e.g., NO_3^-). Total N values for topsoil horizons of most mineral soils range from 0.05 to 0.15%. Under well-aerated conditions, about 1–3% of this organic N will mineralize annually, producing ~15–70 kg N/ha/yr, relative to fertilizer N recommendations for many annual crops of ~50–200 kg N/ha. Long-term use of animal manures or leguminous rotational crops such as alfalfa can greatly increase the amount of potentially mineralizable

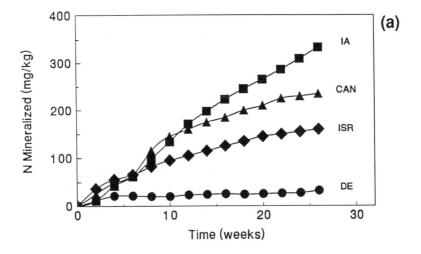

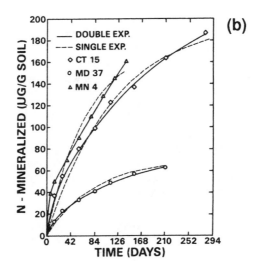

Figure 4-3   (a) Nitrogen mineralization patterns for different soils. (Adapted from laboratory studies conducted by Chae and Tabatabai, 1986 [Iowa, IA]; Ellert and Bettany, 1988 [Canada, CAN]; Hadas et al., 1983 [Israel, ISR]; Sallade and Sims, 1992 [Delaware, DE].) (b) Comparison of single and multiple substrate models to simulate N mineralization in a sludge-amended soil. (Adapted from Deans et al., 1986. With permission.)

organic N in soils. As will be discussed in Section 4.4, this can result in marked reductions in fertilizer N requirements, an important consideration from both economic and environmental perspectives. The timing of N mineralization, relative to the timing of crop N uptake, can be as important a consideration as the amount of N mineralized. Most studies have shown that under optimum conditions, N mineralization follows a curvilinear pattern, as illustrated in Figure 4-3a. The amount of *potentially mineralizable organic N* in soils (N_o) and the rate of N mineralization *(k)* have been successfully described by the use of relatively simple first-order kinetic models that relate the change in mineralized N (N_m) in the soil with time (dN_m/dt) to the amount of mineralizable substrate (N_o) as follows:

$$dN_m / dt = k(N_o) \tag{4-5}$$

Data from laboratory incubation studies that measure the amount of NH_4^+ and NO_3^- leached or extracted from soils at differing time intervals (N_m, t) can be used with the integrated form of this equation to estimate the values of N_o and k for different soil types, or as a function of soil horizonation; or for changing soil chemical and environmental conditions (pH, temperature, moisture); or as affected by long-term changes in soil N caused by differing tillage practices, fertilizers, or organic wastes. Ideally, if N_o and k are known, the amount of mineral N in the soil after a specified time interval could be predicted as:

$$N_m = N_o\left(1 - e^{-kt}\right) \tag{4-6}$$

Some research has shown that a model based on two pools of N_o (N_r = readily decomposable organic N and N_s = slowly decomposable organic N) with separate rate constants for mineralization of each pool (h for N_r, k for N_s) better describes the results of N mineralization studies, particularly for waste-amended soils (Figure 4-3b):

$$N_m = N_r\left(1 - e^{-ht}\right) + N_s\left(1 - e^{-kt}\right) \tag{4-7}$$

Nitrification is the conversion of NH_4^+ into nitrite (NO_2^-) and then NO_3^- by the actions of chemoautotrophic bacteria that are obligately aerobic, i.e., obtain carbon from CO_2 or carbonates and energy from the oxidation of NH_4^+ or NO_2^-. Initially, bacteria of the genera *Nitrosomonas, Nitrosospira,* or *Nitrosococcus* oxidize NH_4^+ to hydroxylamine (NH_2OH); and then, through several other intermediate compounds that are not well-known, to NO_2^-. Two key features of this step are the change in oxidation state of N from -3 to $+3$, and acidification of the soil by hydrogen ions produced when the NH_4^+ is oxidized:

$$2NH_4^+ + 3O_2 \rightarrow NH_2OH \rightarrow 2NO_2^- + 2H_2O + 4H^+ + Energy \tag{4-8}$$

In the next reaction, bacteria of the genera *Nitrobacter, Nitrospira,* or *Nitrococcus* continue the oxidative process, convert NO_2^- to NO_3^-, and change the oxidation state of N from $+3$ to $+5$:

$$2NO_2^- + O_2 \rightarrow 2NO_3^- + Energy \tag{4-9}$$

Nitrate can then be used directly by plants or soil microorganisms, or lost from the crop rooting zone by *denitrification, leaching,* or *erosion/runoff* (Sections 4.2.2 and 4.2.3). In most soils nitrification is a rapid process, somewhat unfortunate given the much greater mobility in soils of NO_3^- than of NH_4^+. Chemical inhibitors of nitrification have been developed to delay this conversion process and have been shown in numerous laboratory studies to be quite effective. To date, however, field research on the effectiveness and economic value of nitrification inhibitors for fertilizers and organic wastes has been inconclusive. The properties and use of some nitrification inhibitors are described in more detail in Section 4.4.2.

Immobilization is essentially the reverse of mineralization, and involves the assimilation of inorganic N (NH_3, NH_4^+, NO_2^-, NO_3^-) by soil microorganisms and the transformation of these mineral forms of N into organic compounds during microbial metabolism and growth. Plant uptake can also be viewed as a form of immobilization, and understanding or controlling the competition between plants and soil microorganisms for inorganic soil N is an important aspect of soil N management. As immobilization represents the formation of organic nitrogenous compounds, it will be controlled to a large extent by the availability of the carbon (C) needed to produce amino acids and proteins. If a large supply of available C is present in the soil, relative to inorganic N, microbial growth and consumption of soluble N will be stimulated, thus enhancing the conversion of soluble N into biomass N. More favorable ratios of available C to available N will result in an excess of NH_4^+ or NO_3^- in the soil, relative to microbial requirements. The carbon:nitrogen ratio (C:N) of native or added organic matter, along with environmental conditions that regulate microbial population growth, will thus control the amount of inorganic N available for plant uptake or other less desirable fates (leaching, denitrification). Stable soil organic matter has C:N ratios ranging from 10:1 to 12:1, while soil organisms have C:N ratios ranging from 5:1 to 8:1. Mineralization of soil organic matter provides adequate C and N for soil microorganisms and, as mentioned above, a small to moderate quantity of available N for plant uptake.

Adding organic amendments with differing C:N ratios to soils, however, can cause significant but reasonably predictable changes in the amount of plant-available inorganic soil N, as illustrated by the results of eight studies on N mineralization from crop residues with C:N ratios ranging from 8:1 to 80:1 (Figure 4-4a). A C:N ratio of ~25:1 (organic matter that is 40% C, 1.6% N) is commonly used as the ratio where mineralization and immobilization are in balance. Adding materials with wide (>30:1) C:N ratios (e.g., straw, sawdust, composted sludge) can cause a rapid increase in microbial biomass and a depletion of available soil N to the point where N deficiency can occur in many plants. Conversely, some organic amendments (e.g., sewage sludge, poultry manure) with very low C:N ratios can produce large excesses of soluble N and must be managed carefully to avoid N losses to sensitive parts of the environment. Composting is an effective means to stabilize the N in rapidly mineralizable organic wastes, as shown in Figure 4-4b. Careful attention to plant N nutrition is required when composts are used, however, as soils amended with composts often exhibit an initial period of N immobilization and then release N more slowly than the rate required by many annual plants. The C:N ratio of added organic matter does not remain constant during the decomposition process, as C from microbial respiration is evolved from the soil as CO_2. With time, therefore, the C:N ratio will decrease into the range where mineralization — not immobilization — is the dominant process, and the soil once again provides some available N for plant uptake.

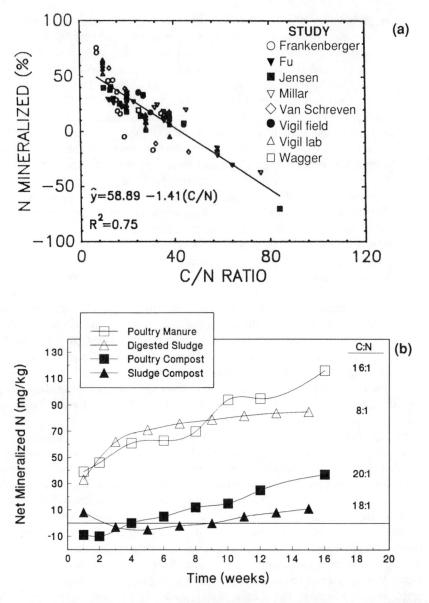

Figure 4-4 (a) Effect of C:N ratio of crop residues on N mineralization. (Adapted from Vigil and Kissel, 1991. With permission.) (b) Effect of composting digested sewage sludge or poultry wastes on N mineralization. (Adapted from Epstein, et al., 1978 and Sims et al., 1992.)

Given the importance of soil microorganisms in mineralization, nitrification, and immobilization reactions, it is apparent that proper management of soil N — particularly from organic sources (crop residues, manures, sludges) — requires an understanding of the soil and environmental factors that can affect the activity of microorganisms controlling these reactions. All parameters that affect biological activity (temperature, moisture, aeration, and soil pH) have been shown to influence the rate and extent of

Table 4-2   Influence of Soil Temperature and Moisture on N
Mineralization

Soil temperature (C°)	Nitrogen mineralization rate constant (k, wk^{-1})	
	Range for 11 soils	Mean
5	0.007–0.015	0.009
15	0.010–0.022	0.014
25	0.019–0.047	0.029
35	0.044–0.071	0.055

Soil moisture tension (Mpa)	Total N mineralized at 35°C (mg/kg)	
	Fine sandy loams	Loams to clay loams
0.01	39	71
0.03	36	67
0.2	29	50
1.5	26	43
Estimated optimum soil moisture content (%)	13	28

Source: Stanford et al., 1973 (temperature); and Stanford and Epstein,
1974 (moisture).

these three N transformations. "Optimum" conditions for each transformation have
been broadly defined and vary slightly between mineralization-immobilization
reactions and nitrification. Because a much wider variety of organisms participate in
mineralization and immobilization, these processes are somewhat less sensitive to
changing environmental conditions than is nitrification. As an example, unlike
nitrification, mineralization can proceed under anaerobic conditions and at much
wider temperature ranges. This can result in an accumulation of NH_4^+ in flooded soils
where nitrification is inhibited by a lack of oxygen, or under extreme soil temperature
regimes (<5°C, >40°C). For mineralization and immobilization, studies have shown
that the optimum conditions are a temperature range of 40–60°C, with a Q_{10} (change
in reaction rate when temperature increases 10°C) of about 2.0, and a soil moisture
content of 50–75% of soil water-holding capacity, although the actual optimum
moisture percentage varies with soil texture (Table 4-2). For nitrification, optimum
conditions include temperatures of 30–35°C, a moisture content of 50–67% of soil
water-holding capacity, and a pH between 6.6 and 8.0. Nitrifying organisms are more
sensitive to excessive soil acidity, and their activity decreases markedly when the soil
pH is less than 5.0. Nitrate production has been observed in highly acidic mine soils
and forest soils, suggesting that some nitrifiers have adapted to these unfavorable pH
conditions.

4.2.2 Gaseous Losses of Nitrogen: Ammonia Volatilization and Denitrification

Ammonia volatilization refers to the loss of NH_3 from the soil as a gas and is
normally associated with high free NH_3 concentrations in the soil solution and high
soil pH. Surface applications of ammoniacal fertilizers or readily decomposable
organic wastes (e.g., manures, sludges) to soils can result in considerable N loss by
NH_3 volatilization, particularly if the soil (or organic waste) is alkaline in nature. This

reaction and the major factors influencing the magnitude and rate $(d[NH_3]/dt)$ of N loss by volatilization can be summarized as follows:

$$NH_4^+ \leftrightarrow \left(NH_3\right)_{solution} + H^+ \qquad (4\text{-}10)$$

$$\left(NH_3\right)_{solution} \leftrightarrow \left(NH_3\right)_{gas} \qquad (4\text{-}11)$$

and

$$\frac{d[NH_3]}{dt} = [A] \times [K] \times \left[P_l - P_g\right] \qquad (4\text{-}12)$$

where
$d[NH_3]/t$ = NH_3 loss in time (t)
A   = Area of soil-solution interface
K   = Mass transfer coefficient, a function of air velocity above the soil and the air and soil temperatures
P_l = Partial pressure of NH_3 in soil solution
P_g = Partial pressure of NH_3 in air above soil solution

Volatilization represents both the loss of a plant nutrient and a potential environmental impact of N, as studies have shown that nearby surface waters can be enriched by NH_3 volatilized from areas where organic wastes are concentrated (e.g., feedlots, manure lagoons). As described above, the key factors in the volatilization of NH_3 from soils are those that affect: (1) the transfer of a gas between the soil solution and the atmosphere (area of solution-atmosphere interface, velocity of air across the soil surface) and (2) the general rate of a chemical reaction (temperature, partial pressure of NH_3 in both phases). Soil and management factors that control these two aspects of NH_3 volatilization include soil temperature, moisture, and texture; nature of N source; and methods of application of fertilizers or organic wastes (e.g., surface broadcast, injected, incorporated). Conditions associated with maximum volatilization losses of NH_3 will include surface applications of fertilizers or manures, high pH or calcareous soils, soils with low cation exchange capacities and therefore little ability to retain the NH_4^+ cation, and a warm, slightly moist environment. In general, the most effective method to reduce NH_3 volatilization losses is by incorporating the N source into the soil by tillage, injection, or irrigation — through timing of application or by natural rainfall (as shown in Figure 4-5 and Table 4-3). Once incorporated, the cationic nature of NH_4^+ results in its electrostatic attraction to cation exchange sites on clays and organic matter, reducing the likelihood of NH_3 loss. Volatilization is a particular problem for pastures, turf, and no-tillage agriculture where surface applications of fertilizers and manures are required.

Urea $[CO(NH_2)]$ fertilizers and uric acid found in manures and other organic wastes represent a special case of NH_3 volatilization. When urea is added to a soil, it is decomposed in a reaction that is catalyzed by the enzyme urease (urea amidohydrolase), as shown below:

$$CO\left(NH_2\right)_2 + H_2O \xrightarrow{\text{Urease}} \left(NH_4\right)_2 CO_3 \text{(Unstable)} \rightarrow CO_2 + 2NH_3 \qquad (4\text{-}13)$$

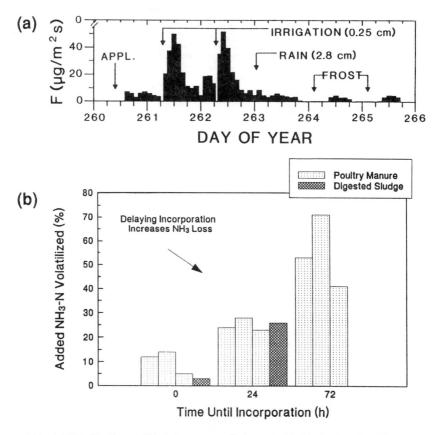

Figure 4-5 (a) Volatilization of NH_3 from urea applied to a moist, bare soil surface. Low rate of irrigation (0.25 cm) dissolves urea, initiates NH_3 loss (F). Rainfall (2.8 cm) leaches NH_3 into soil, volatilization rate decreases markedly. (Adapted from McInnes et al., 1986. With permission.) (b) Effect of time between application of poultry manure and digested sewage sludge on NH_3 losses from soils. (Adapted from Donovan and Logan, 1983; Gartley and Sims, 1993.)

Urease is produced by soil microorganisms involved in the decomposition of soil organic matter, in crop residues, and in other organic materials. Urease acts by rapidly hydrolyzing the C–NH_2 bonds in the urea molecule. The rapid formation of a highly concentrated solution of NH_4^+ in the microenvironment surrounding adjacent urea granules increases the pH to about 8.5, greatly enhancing volatilization of NH_3. Chemical inhibitors of the urease enzyme have been developed to reduce the rate of urea hydrolysis and thus the potential for N losses by NH_3 volatilization.

Denitrification is defined as the reduction of NO_3^- to gaseous forms of N (NO, N_2O, N_2) by chemoautotrophic bacteria, as follows:

$$NO_3^- \rightarrow NO_2^- \rightarrow NO \rightarrow N_2O \rightarrow N_2 \qquad (4\text{-}14)$$

nitrate   nitrite   nitric   nitrous   dinitrogen
                     oxide   oxide

The bacteria responsible for denitrification are normally aerobic, but under anaerobic conditions can use NO_3^- as an alternative to oxygen (O_2) as an acceptor of electrons

Table 4-3   Effect of Application Method and Rainfall on Corn Yields Where Volatile NH_3 Fertilizers (UAN, Urea) Were Used

N Source	Application method	Application method N rate (kg/ha)			Mean
		90	180	270	
		Corn yield (Mg/ha)			
UAN	Broadcast spray	5.6	6.8	7.2	6.5
	Surface band	7.4	8.3	8.7	8.1
	Incorporated band	7.9	8.8	8.7	8.5

N Source	Rainfall	Rainfall effects N rate (kg/ha)			Mean
		50	101	202	
		Corn yield (Mg/ha)			
NH_4NO_3	25 mm within	10.1	11.3	11.6	11.0
Urea	36 hr of application	10.4	11.7	11.2	11.1
NH_4NO_3	None for 3 d then	8.8	10.7	11.1	10.2
Urea	3 mm	8.5	9.2	10.4	9.4

Source: Touchton and Hargove, 1982 (application method); and Fox and Hoffman, 1981 (rainfall effects).

Note: In rainfall effects study, NH_4NO_3 represents a stable N source with low volatilization potential.

produced during organic matter decomposition. During the denitrification process, N is reduced from an oxidation state of +5 in N_3^+ to 0 in N_2. The critical factors regulating the rate and duration of denitrification in soils are the availability of NO_3^- (the substrate) and C (source of energy and electrons) and the absence of O_2. For denitrification to occur, the soil must first produce or be amended with NO_3^- and then enter an anaerobic period. Conversely, denitrification can be inhibited by any process that restricts NO_3^- production (e.g., slows nitrification), enhances NO_3^- removal (leaching, plant uptake), or promotes aerobic conditions (artificial drainage, plant depletion of soil moisture).

As NO_3^- originates primarily by nitrification, an aerobic process, the conditions most conducive to denitrification are those that involve alternating aerobic and anaerobic cycles or adjacent aerobic and anaerobic zones. Soils that are periodically flooded or experience transitory anaerobic conditions due to heavy rains, can accumulate pools of NO_3^- during an aerobic period that is lost by denitrification during the subsequent anaerobic cycle. Rapid O_2 consumption by soil microorganisms during the decomposition of organic matter can produce anaerobic zones adjacent to aerobic areas where NO_3^- is being produced. When NO_3^- moves into the anaerobic zone by diffusion or mass flow, it can then be denitrified, as illustrated in Figure 4-6 for a soil receiving slit injections of a pharmaceutical waste. Common examples of these anaerobic zones are microsites in the rhizosphere that have been enriched with root exudates, localized deposits of highly decomposable crop residues, and organic wastes injected into relatively small soil volumes, as opposed to a more uniform spatial distribution. An ecological example would be a wetland area adjacent to an agricultural

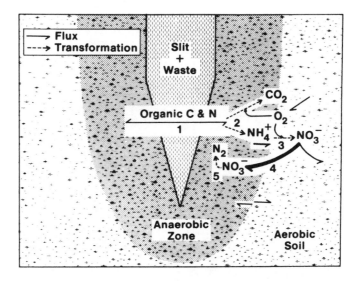

Figure 4-6 Denitrification induced by slit injection of a highly decomposable pharmaceutical waste. Concentrating available carbon stimulates high oxygen consumption, producing an anaerobic zone adjacent to an aerobic zone where nitrification occurs as waste mineralizes. (Adapted from Rice et al., 1988. With permission.)

field or perhaps an artificial wetland used to treat wastewaters (Figure 4-7). In this situation, a large pool of NO_3^- produced in the normally aerobic field soil moves by leaching and lateral flow into a wetland ecosystem dominated by flooded soils, anaerobic conditions, and high quantities of available C, and is then removed from the drainage waters by denitrification.

Soil temperature and pH can also influence denitrification. Although denitrification can occur at temperatures between 2 and 75°C, the optimum temperature is ~30°C. Soil temperature affects not only the rate of microbial metabolism, but also influences chemical processes such as the rate of diffusion of O_2, NO_3^-, N_2O, and N_2 in the soil water and atmosphere. The optimum pH for denitrifying organisms ranges from 6.0 to 8.0, but as with nitrification, denitrification has also been measured in highly acidic soils.

4.2.3 Leaching and Erosional Losses of Nitrogen

Nitrogen can be transported from soils into groundwaters and surface waters by *leaching, erosion,* or *runoff.* Losses of N by leaching occur mainly as NO_3^- because of the low capacity of most soils to retain anions. In general, any downward movement of water through the soil profile will cause the leaching of NO_3^-, with the magnitude of the N loss proportional to the concentration of NO_3^- in the soil solution and the volume of leaching water. Leaching of NO_3^- is economically and environmentally undesirable. Nitrate that leaches below the crop rooting zone represents the loss of a valuable plant nutrient and as mentioned earlier can contribute to pollution of groundwater aquifers and eutrophication of surface waters, as shown in Figure 4-8. Much of the research conducted with fertilizers, animal manures, and other by-products

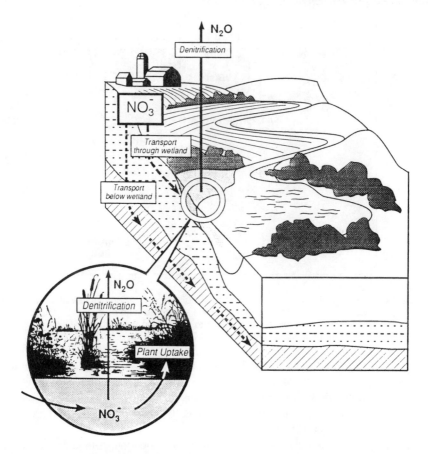

Figure 4-7   Wetlands as denitrifying zones capable of removing NO_3^- from runoff or subsurface drainage from agricultural fields.

(e.g., sewage sludges) has been directed toward reducing NO_3^- leaching, especially in humid regions. Nitrate pollution of groundwaters, however, is not a universal problem and is often regional or local in nature, as shown in Figure 4-9.

Situations most conducive to NO_3^- leaching and groundwater pollution include sandy, well-drained soils with shallow water tables, in areas that receive high rainfall or intensive irrigation and frequent use of fertilizers, manures, or other N sources. Nitrate leaching concerns are not restricted to these situations, however. Any situation involving overapplication of N, organic waste storage areas (e.g., feedlots, lagoons), or intensive irrigation has the potential to cause significant NO_3^- leaching, regardless of soil type and climate. Chemical retention of NO_3^- in soil profiles — unlike other anions (phosphate, sulfate) — is generally of little value in reducing leaching, although some highly acidic subsoils have been shown to have significant anion exchange capacity. Denitrification in groundwaters or groundwater discharge areas (e.g., wetlands) may reduce leaching losses of N. Subsoil denitrification, however, is likely to be of little value in mitigating NO_3^- leaching, primarily because of the low available C levels in most subsoils. In any case, given the atmospheric

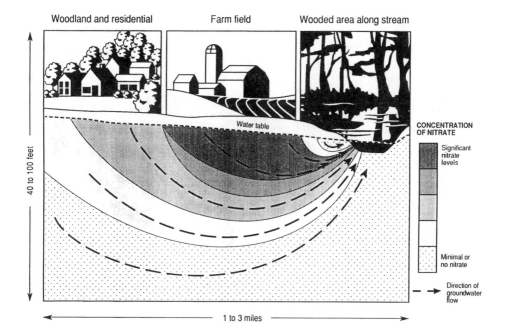

Figure 4-8 Generalized representation of NO_3^- transport in groundwater aquifers, illustrating variable nature of NO_3^- concentration with land use, aquifer depth, and distance to discharge area. (Adapted from Hamilton and Shedlock, 1992.)

impacts of N oxides, management techniques designed to control leaching by enhancing denitrification are certain to be examined carefully. Approaches to minimize NO_3^- leaching, therefore, are operational in nature and focus on controlling the timing of NO_3^- formation in soils, and understanding the soil and climatic conditions of the region and the N uptake patterns of the dominant crops grown. These approaches, their implications, and constraints are discussed in more detail in Section 4.4 of this chapter.

Erosion refers to the transport of soil from a field by water or wind; *surface runoff* is the water lost from a field when the rate of precipitation exceeds the infiltration capacity of the soil. Both processes can transport soluble inorganic N and organic N to surface waters and contribute to the process of eutrophication or drinking water contamination. Many watershed studies have shown that most of the N lost by erosion or runoff is sediment-bound organic N. Although the solubility of NO_3^- favors its loss in runoff as opposed to sediment transport, total N losses from most watershed studies are usually severalfold greater than soluble N.

Surface applications of organic wastes are undesirable because they increase the likelihood of soluble and organic N losses by erosion and runoff. This approach is usually not permitted with municipal and industrial wastes; however, in agricultural operations, conservation practices designed to control erosion by reducing tillage can allow the application of animal manures to soil surfaces. Surface applications of

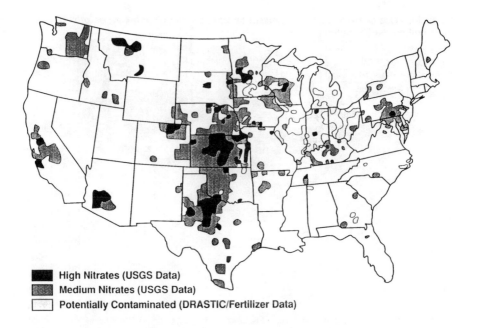

Figure 4-9   Potential for groundwater contamination by NO_3^- in the U.S., illustrating regional nature of the problem. (Adapted from Nielsen and Lee, 1987.)

manures can also occur when farmers apply manures during winter months, when the soil is frozen and less susceptible to equipment damage and erosion and when more time is available. The use of grassed waterways or border strips that trap sediment and accumulate soluble N in plant biomass can help reduce N losses in these situations, as shown in Table 4-4 where a cornstalk residue strip 2.7 m wide with 50% ground cover reduced sediment and total N losses by 70–80%.

4.2.4 Biological N Fixation

Biological N fixation is the conversion of atmospheric N_2 into an organic form of N, either through symbiotic associations between plants and microorganisms or independently by free-living organisms such as cyanobacteria ("blue-green algae") and certain heterotrophic bacteria. In the context of the global N cycle, biological N fixation represents the major N input to soils.

Symbiotic N fixation can occur between leguminous plants and bacteria, between nonlegumes and actinomycetes, and in some plant-algal associations. The general characteristic of this symbiosis is the ability of the N-fixing organism to enzymatically reduce N_2 to NH_3 when provided with an energy supply (photosynthate) by the host. From an agricultural perspective, the most important type of symbiotic N fixation is that occurring between leguminous plants and bacteria of the genera *Rhizobium* and *Bradyrhizobium*. Although we will focus on N fixation by legumes in this chapter, it should be recognized that the ability of nonlegumes to fix atmospheric N_2 can be equally important to nonagricultural ecosystems and to certain crops. An example of nonleguminous, N fixation is the symbiotic relationship

Table 4-4 Use of Residue Strips to Reduce Erosion and Runoff Losses of N from a Bare Soil with a 5% Slope

Strip width, residue cover, and antecedent moisture	Entering residue strip			Leaving residue strip		
	Sediment[a]	Runoff[a]	Total N[b]	Sediment[a]	Runoff[a]	Total N[b]
1.8 m and 27% Cover						
Dry	9.5	171	22.5	6.0	244	16.5
Very wet	16.6	320	31.9	14.3	373	28.2
2.7 m and 50% Cover						
Dry	22.7	284	45.3	4.9	280	14.1
Very wet	24.0	386	39.4	4.9	462	13.0

Source: Alberts et al., 1981.
[a] kg/hr/m of width.
[b] g/hr/m of width.

between actinomycetes of the genus *Frankia* and a wide variety of trees and woody shrubs that are important in soil formation, in revegetation of disturbed, highly erodible forest soils, and as sources of fuelwood. A nonleguminous, agricultural example of symbiotic N fixation is the association between the cyanobacterium *Anabaena* and the freshwater, free-floating fern, *Azolla,* found widely in flooded, tropical areas used for rice production. *Anabaena* are located on the stem and fronds of *Azolla,* but unlike *Rhizobium* or *Frankia,* they do not directly convert N_2 into organic N within the plant. Instead *Anabaena* produce soluble NH_4^+ that is then taken up by *Azolla* roots growing in the water or upper portion of the sediment of a rice paddy. This fixed N can then be used directly by rice when the *Azolla* die and the organic N is mineralized; or in some countries the *Azolla* is harvested and used as animal feed or organic fertilizer.

Leguminous plants have been important components of agricultural crop production systems for centuries, and represent a major pathway for the conversion of atmospheric N to soil N and then to a form that can be used by humans and animals. Over 14,000 species of legumes are known, of which more than 100 are commonly used in agriculture. In addition to their direct value as food crops, some legumes are grown solely as cover crops to provide N for a subsequent nonleguminous crop (e.g., clovers or vetches followed by corn). Environmental issues related to legumes will be discussed in more detail in Section 4.4.2, but are similar to those associated with other organic N sources, and primarily center on the management of legume N to avoid NO_3^- leaching, runoff, and denitrification.

The process of symbiotic N fixation by legumes begins when bacteria living in the soil near the root system of a host plant attach themselves to cells of root hairs and in doing so induce a curling of the root hair around the bacteria. The bacteria then produce an infection thread that invades the root cell, allowing bacterial penetration and proliferation within the root cortex. In response to this bacterial invasion, the plant synthesizes a nodule, a protective structure that encapsulates the bacteria which have now changed both physical shape and metabolic function and are now referred to as bacteroids (Figure 4-10). The bacteria-host plant relationship is often quite specific, and a given species of *Rhizobium* or *Bradyrhizobium* will not infect or nodulate incompatible host plants. The nodule is comparable to other energy-converting organs in the plant, and possesses a membrane that regulates the entry and exit

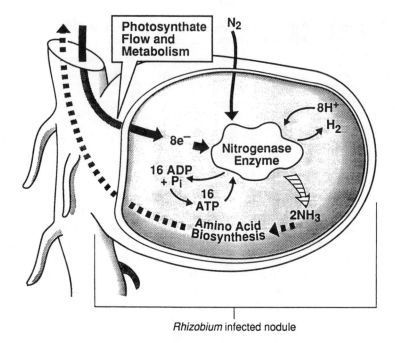

Figure 4-10 Schematic representation of the form and function of bacteroids in the nodules of legume roots.

of metabolites to and from the bacteroids. In the N-fixation reaction the bacteroids use photosynthate-derived energy (ATP and electrons), oxygen, and a specific enzyme, *nitrogenase,* to reduce N_2 to NH_4^+ which is then released into the host cell and used in amino acid synthesis. The overall equation for N fixation is:

$$N_2 + 16ATP + 8e^- + 10H^+ \xrightarrow{\text{Nitrogenase}} 2NH_4^+ + H_2 + 16ADP + 16P_i \quad (4\text{-}15)$$

Each host plant-bacteria association has its own genetic potential for N fixation. For a given symbiosis, the actual amount of N fixed depends primarily on available soil N and the energy status of the plant; this in turn is influenced by such factors as carbohydrate supply, light intensity, soil nutrient status, soil temperature, and moisture. Examples of agriculturally important legumes and estimated amounts of N fixed under optimum conditions are given in Table 4-5.

A generalization of environmental factors important in N fixation would include soil pH (near neutral), supply of key nutrients (P, K, Ca, Fe, Mo, Co), "optimum" temperature and moisture (varies with host plant), adequate aeration, and a low level of available soil N (NH_4^+ and NO_3^-). It is important to remember, however, that even under optimum conditions legumes do not obtain all of their N from biological N fixation. Soil N can provide as much as 50% of the total N in legumes with low fixation rates. Also, if the concentration of NO_3^- in the soil solution exceeds ~1 mM, the process of N fixation is restricted. This can occur when legumes are planted in soils with high levels of residual N, perhaps resulting from previous fertilizer or organic waste applications. If supplied with sufficient N fertilizer, legumes will not

Table 4-5 Major Legume Crops Used in Agriculture and
Estimates of Annual N Fixation

Crop and N-fixing bacteria		Annual N$_2$ fixation	
		Range (kg/ha/yr)	Typical value (kg/ha/yr)
Host plant	*Rhizobium*		
Alfalfa	*R. meliloti*	60–500	225
Clovers	*R. trifolii*	60–350	115
Peas, vetch	*R. leguminosarum*	90–180	100
Beans	*R. phaseoli*	20–100	45
Host plant	*Bradyrhizobium*		
Lupins	*B. lupinii*	150–170	160
Soybeans	*B. japonicum*	65–200	100
Cowpeas	*B. parasponiae*	65–130	100

Note: *Rhizobium* are fast growing symbiotic bacteria; *Bradyrhizobium* are slow growing.

nodulate and N fixation will not occur. This has led to debates over the economic value of "starter" N fertilizers for legumes (e.g., soybeans) as well as the environmental impact of applying organic N sources, such as manure and sludges, to legumes. The presence of a small amount of available N prior to effective nodulation may benefit the plant, but excessive fertilizer N will often reduce or stop biological N fixation.

In addition to symbiotic processes, N fixation can occur with some species of "free-living" organisms. The major examples are the cyanobacteria (e.g., *Anabaena,* see above) that are autotrophic (requiring only light, water, N$_2$, CO$_2$, and salts) and certain heterotrophic bacteria. Important nonsymbiotic, N-fixing bacteria include *Azotobacter* and *Beijerinckia,* aerobic saprophytes that obtain their energy from the oxidation of organic matter; *Azospirillum,* facultatively anaerobic bacteria; and *Clostridium,* anaerobic saprophytes. In general, because the amount of N fixed by these bacteria is small (5–50 kg per N/ha), they are of little importance to most soil N management programs.

4.3 SOURCES OF NITROGEN

The production of food and fiber, the growth of plants for aesthetic purposes, and the reclamation and stabilization of lands disturbed by construction, mining, and other industrial or urban activities, often requires the addition of supplemental N to obtain optimum plant growth. Two broad categories of N sources exist, inorganic and organic. Inorganic N sources are predominantly commercial fertilizers, but also include limited quantities of mineral deposits and industrial by-products. Organic N sources commonly include animal manures, crop residues, municipal sewage sludges and wastewaters, and a wide variety of industrial organic wastes. For most of recorded history the major source of N added to soils was organic in nature, primarily as animal manures or crop residues. The first N fertilizer used commercially in the United States, Peruvian guano, was also organic in nature; it was formed from centuries of deposition of excreta by seafowl along the South American coast. In the late 1800s and early 1900s, industrial processes were devised that could fix atmospheric N_2, converting it to NH_3 gas, calcium cyanamide, or nitric acid. Later improvements in the efficiency of one such process — the Haber-Bosch method for the synthesis of NH_3 gas from atmospheric N_2 — resulted in the availability of inexpensive, high analysis N fertilizer materials, and began to markedly change the nature of agriculture on a global scale. Food production was no longer limited by the availability of soil, manure, or legume N because fertilizer N could now be used to greatly increase the yield per hectare of most agricultural crops. Beyond this, the ease of handling of commercial N fertilizers reduced the labor requirements of crop production and contributed to the development of larger, more specialized farms devoted in many cases to the production of fewer crops.

Increased production capacity did not always equate to increased efficiency of N use, however. As farmers (and those advising farmers) learned to use new fertilizer materials alone or in combination with organic N sources, new application equipment, and new cropping systems, they often proceeded without a real understanding of the potential environmental impacts involved. The uncertainty associated with these new production practices and the low costs of fertilizer N undoubtedly resulted in overfertilization with N and contributed to groundwater contamination by NO_3^-, especially in areas of intense fertilizer use.

Regardless of the form of N used or the nature of the soil-plant system, maximizing the efficiency of N recovery and minimizing the potential of the N source to pollute the environment are now fundamental goals of modern agriculture. To properly manage either type of N source, it is essential to understand how the physical, chemical, or biological properties of the material affect its handling, application, and fate among the various transformations of the soil N cycle. It is also important to understand that all environmental problems caused by improper N management can occur with both inorganic and organic N sources. That is, despite the great interest in sustainable agriculture and organic farming, research has found no intrinsic superiority associated with organic N sources. In fact, in many situations, the physical properties and heterogeneity in composition of organic wastes can make them much more difficult to manage successfully than commercial N fertilizers.

The purpose of the following sections is to provide an overview of the production, composition, and characteristics of the major inorganic and organic sources of N used as soil amendments and to illustrate key aspects related to sound environmental use of all N sources.

4.3.1 Inorganic Sources of Nitrogen

Most commonly used N fertilizers, summarized in Table 4-6, are produced from NH_3 gas synthesized by the Haber-Bosch process. This process uses natural gas (CH_4), atmospheric N_2, and steam (H_2O) to produce NH_3 gas as follows:

$$7CH_4 + 10H_2O + 8N_2 + 2O_2 \rightarrow 16NH_3 + 7CO_2 \qquad (4\text{-}16)$$

This clearly illustrates another environmental aspect of N use, its impact on natural resources, because the production of N fertilizers consumes natural gas, a finite and critically important natural resource. Efficient use of N fertilizers will thus enhance the longevity of natural gas supplies.

Table 4-6 Properties of Major Commercial Nitrogen Fertilizers

Nitrogen source	Chemical composition	%N
Ammoniacal N sources		
Anhydrous ammonia	NH_3	82
Aqua ammonia	$NH_3 \cdot H_2O$	20–25
Ammonium chloride	NH_4Cl	25
Ammonium nitrate	NH_4NO_3	33
Ammonium sulfate	$(NH_4)_2SO_4$	21
Nitrate N sources		
Calcium nitrate	$Ca(NO_3)_2$	15
Potassium nitrate	KNO_3	13
Sodium nitrate	$NaNO_3$	16
Urea materials		
Urea	$CO(NH_2)_2$	45
Urea-ammonium-nitrate solutions	30–35% Urea:40–43% NH_4NO_3	28–32
Ureaform	Urea-formaldehyde	38
IBDU	Isobutylidene diurea	32
SCU	Sulfur-coated urea	36–38
Nitrogen-phosphorus materials		
Monoammonium phosphate (MAP)	$NH_4H_2PO_4$	11
Diammonium phosphate (DAP)	$(NH_4)_2HPO_4$	18–21
Ammonium polyphosphates (liquid)	$(NH_4)_3HP_2O_7$	10–11

Once NH_3 has been synthesized it can be used: (1) directly as anhydrous ammonia, a pressurized gas [NH_3]; (2) reacted with CO_2 to form urea [$CO(NH_2)_2$]; (3) oxidized to NO_3 and reacted with more NH_3 to form ammonium nitrate [NH_4NO_3]; and, (4) combined with sulfuric acid to produce ammonium sulfate [$(NH_4)_2SO_4$] or with various types of phosphoric acid to form ammonium phosphates such as diammonium phosphate [DAP, $(NH_4)_2HPO_4$] or monoammonium phosphate [MAP, $NH_4H_2PO_4$]. Further industrial processes can be used to produce N solutions (such as urea-ammonium-nitrate [UAN] or aqua ammonia [NH_3]) or controlled, slow-release, solid N fertilizers that are coated with resins (Osmocote, used in greenhouses and nurseries) or sulfur (sulfur-coated urea) to delay their rate of dissolution in the soil. A wide variety of mixed fertilizers containing N, P, K, and other nutrients are also produced. Many urban situations (e.g., turf, home gardens) and land reclamation projects make extensive use of mixed fertilizers to provide N and enhance the overall nutritional status of low-fertility soils. Nitrate-N fertilizers (such as calcium or potassium nitrate [$Ca(NO_3)_2$, KNO_3]) are also available; however, due to the production costs and lower efficiency of N recovery from NO_3-based materials, they are primarily used on specialty crops (vegetables, fruits, tobacco) or on crops that are highly sensitive to NH_3.

Economics, more than any other factor, often controls the N source selected. In general, fertilizers with higher N contents have lower costs of storage, transportation, handling, and application; and hence are more economical. However, the properties of some high analysis N fertilizers can increase these costs. For example, anhydrous ammonia, which has the highest N content of any fertilizer material (82% N), requires complex application equipment that must pressurize the gas, convert it to a liquid, and inject the NH_3 into the soil in a manner that reduces volatilization losses. Ammonia volatilization is also a major concern with urea, the highest analysis solid N fertilizer (45% N); and some type of incorporation is often required to maximize the efficiency of N recovery from this N source. The physical and chemical properties of N sources are not the only factors that influence source selection, however. In the United States there are marked regional preferences for N fertilizers, often related more to the fertilizer manufacturing and transportation infrastructures that have evolved with time than to specific requirements of regional cropping systems (Figure 4-11).

Other factors that influence the selection of a N fertilizer include crop management practices (crop type and rotation, tillage, irrigation), the soil type and sensitivity to N losses, the effect of the N source on soil pH (nitrification is an acidifying process and N fertilizers vary in their potential to decrease soil pH), the need to simultaneously supply other nutrients, and the suitability of the material for existing application equipment. The implications of these factors for N management are discussed in more detail in Section 4.4.

4.3.2 Organic Sources of Nitrogen

A wide variety of organic materials are used as soil amendments, many of which contain appreciable quantities of N. Animal manures, municipal sewage sludges and wastewaters, composts of municipal solid waste or yard waste, food processing wastes, industrial organics, and crop and forest product residues are the dominant organic wastes produced worldwide. Estimates of the quantity of organic wastes produced annually in the United States exceed 400 million mt/yr. The magnitude of this problem is even more apparent when expressed on a "per-person" basis. For animal manures, municipal solid wastes and sewage sludges, industrial solid wastes and water treatment sludges, and silviculture residues, the amount generated is roughly equivalent to 900, 550, 450, and 400 kg per person per year (dry weight basis), respectively. Changes occurring in the U.S. and other countries are likely to alter the nature and distribution of organic wastes significantly. In many states landfilling of yard wastes (leaves, grass clippings), which accounted for ~20% of the landfill volume, is rapidly becoming an unacceptable practice and is being replaced with the commercial production of yard waste composts. Similarly, a number of municipalities have developed composting facilities for sewage sludge and the organic fraction of municipal refuse. Production of these composts is certain to result in large increases in the amount of organic N applied to soils. Other changes occurring include the production of composted or pelletized animal manures in areas where large excesses of manure, relative to arable land, exist. Movement of animal manures off the farm for use in urban areas, or in construction and reclamation projects will likely result in an increased use of these materials in land application programs involving horticultural crops, turf grasses, and revegetation of sites for use by domestic animals and wildlife.

Although all organic wastes contain N, the amount, forms, and availability of N can vary widely. Mean N contents and ranges are available for most organic wastes, but the considerable variability in total and inorganic (e.g., NH_3, NO_3^-) N contents of most wastes makes interpretation of analytical results for N or other elements an ongoing problem. Broad generalizations of the N content in organic wastes are probably justified because it has been adequately documented that certain organic wastes will consistently have higher total N contents than others, as shown in Table 4-7 for animal manures (poultry > swine > dairy > beef) and municipal sludge

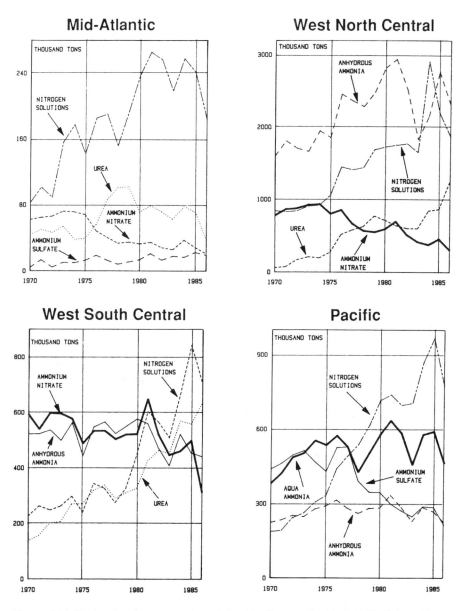

Figure 4-11 Regional preferences and trends in N fertilizer use for 1970–1985. (Adapted from Berry and Hargett, 1986.)

products (aerobic > anaerobic > composted). The wide range in total N content among similar types of wastes, however, can have significant implications for N loading to soils and crops, as shown in Figure 4-12. In this study, the amount of manure needed to provide a desired amount of N for corn was estimated for poultry manure based on analysis of manure samples from 17 different on-farm storage areas. When the predicted amount of N to be added was compared to the actual N applied, based on analysis of manure samples collected during application, overapplication of

Table 4-7 Representative Values for N Content and Availability for
Selected Organic Wastes

Organic N source	Total N (%)	Organic N Mineralized[a] (%)
Animal manures		
Beef	1.3–1.8	25–35
Dairy	2.5–3.0	25–40
Poultry	4.0–6.0	50–70
Swine	3.5–4.5	30–50
Sludge products		
Aerobic digestion	3.5–5.0	25–40
Anaerobic digestion	1.8–2.5	10–20
Composted	0.5–1.5	(−10)–10
Other wastes		
Fermentation wastes	3.0–8.0	20–50
Poultry processing wastes	4.0–8.0	40–60
Paper mill sludges	0.2–1.0	(−20)–5

Note: Average values from various sources.
[a] Organic N mineralized estimated from laboratory incubation studies. Negative values for composts and paper mill sludges indicate that immobilization of N occurred.

10–20 kg N/Mg of manure commonly occurred, as did underapplication of 5–10 kg N/Mg. Therefore, the accurate application of a recommended manure rate for corn (~5 Mg/ha), based on analysis of the manure, commonly resulted in the application of *excess* manure N approaching the total N requirement of the crop (~100 kg N/ha). Clearly, a more comprehensive approach than N analysis and equipment calibration is needed to avoid over- or underapplication of N from organic wastes.

Some recent and innovative approaches to organic N management are described in Section 4.4 of this chapter. In general, however, the key to effective use of organic N sources is an understanding of the factors that influence the extent and rate of conversion of organic N to forms that are available for plant uptake or loss to the environment. The availability of N in organic wastes will be influenced by their composition, largely controlled by waste production and storage practices, and by the chemical and biological changes they undergo following application to the soil. Much of the research in this area has been directed toward identifying the differences in N availability between various types of wastes and then using waste properties to predict N availability. As with waste composition, these studies have shown that wastes can be broadly grouped in terms of organic N availability (Table 4-7). Certain wastes (e.g., poultry manure, aerobically digested sludges) not only are higher in total N but also will provide more N on decomposition in the soil than other more stable wastes (e.g., composts, paper mill sludges). Simple and complex approaches have been taken to predict N availability. Sewage sludges have been ranked according to N mineralization potential as waste-activated (40%) > raw and primary (25%) > anaerobically digested (15%) > composted (8%). For some wastes, both total and NH_4-N must be included to accurately estimate the amount of potentially available N (PAN), as shown in a simple model developed for poultry manures:

$$PAN = \left[k_m (N_o) + e_f (NH_4\text{-}N) \right] \tag{4-17}$$

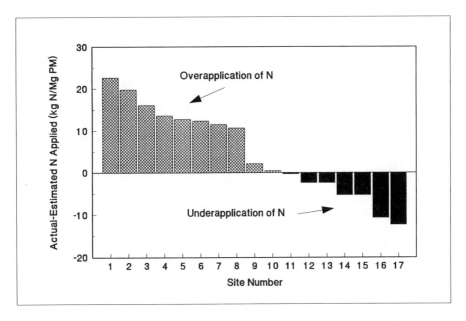

Figure 4-12 The difference between total N based on poultry manure (PM) samples collected during field application and the amount estimated to be applied based on laboratory analyses of stockpiled manure samples. Results of 17 field studies. (Adapted from Igo et al., 1991.)

where

k_m = percentage of organic N (N_o) mineralized (K_m = 40–60%, depending on season of year manure was applied, e.g., winter vs spring)

e_f = factor reflecting the efficiency of recovery of manure NH_3-N (20–80%, depending on time until incorporation of manure)

The use of organic wastes as N sources, however, cannot be based on N availability alone. Other nutrients, or nonessential elements, in these materials can determine not only the application rate but also their suitability for various end uses. At present, there are three main aspects of waste composition that affect both short- and long-term use of organic wastes in land application programs: (1) the P buildup to excessive levels in waste-amended soils; (2) the potential for heavy metal contamination of soils, crops, and waters; and (3) the presence and possible adverse environmental impacts of organic pollutants found in organic wastes. These factors will be discussed in greater detail in Chapter 9.

4.4 PRINCIPLES OF EFFICIENT NITROGEN MANAGEMENT

Efficiency can be defined as the ability to accomplish a task without the waste of time, energy, or resources. Nitrogen efficiency for a soil-plant system, in its simplest form, can be viewed as the ability to manage the time, energy, and resources needed to obtain an acceptable level of plant growth (the task) with minimal loss of N (the waste). Few, if any, natural systems are 100% efficient; and given the complexity of the soil N cycle and the constraints imposed by time, labor, soil type, cropping practices, environmental conditions, and available resources, it is perhaps not surprising to find that efficiency values for the recovery of applied N in most cropping systems rarely exceed 60% and commonly range between 30 and 50% under normal management. Unfortunately where N is concerned, inefficiency of recovery by one ecosystem (soil-plant) often results in the redistribution of NO_3^-, NH_3, or N_2O to another. This represents a second impact of N efficiency, the potential for adverse effects on another resource, the environment. Based on the above definition of efficiency, the key questions that must be addressed for N management are:

- How do we define an acceptable level of plant growth, particularly in nonagricultural systems such as urban areas, forests, and land reclamation sites?
- How do we decide what degree of N "waste" is unacceptable, given the complex interactions between the politics and economics of plant production; the often unpredictable cycling of N between soil, air, and water; and the varying degrees of ecosystem sensitivity to N pollution?
- Perhaps most important, how do we design more efficient systems for N use, and monitor them well enough to know whether they are improvements over past practices?

Questions such as these rarely have straightforward answers; instead we respond to them by the development of processes and management systems that continually evolve as our knowledge base grows. Acceptable levels of plant production must consider both global food shortages and the profit margins of individual farmers. In many urban areas and areas dominated by animal-based agriculture, they must also reflect urgent waste disposal needs caused by the concentration of nutrients in areas that do not have the land base to use them. If a soil has been amended with wastes for years and requires little if any N to produce the maximum yield attainable given other inflexible constraints (e.g., sunlight, rainfall), how do we reconcile the concept of acceptable plant growth with the continued generation of manure or sludge N? Similarly, environmental regulatory agencies have established unacceptable levels of N in drinking waters and surface waters; however, in some areas research has shown that — due to the nature of the soils, climate, and limitations imposed by current technology — we cannot profitably produce grain crops without having drainage waters that exceed these standards. Can we improve efficiency enough by management or legislation to overcome the fundamental limitations on crop N recovery in these areas? Even more important, how will we determine whether our new, and perhaps more expensive, practices are significantly reducing the N pollution of groundwaters, surface waters, or the atmosphere?

Complex problems such as these require multidisciplinary management approaches. The central challenge faced at present by research, advisory, and regulatory agencies worldwide is the need to integrate the expertise of many disciplines to develop management plans that maximize the economic value of the N contained in fertilizers and organic wastes, while minimizing the adverse environmental impacts. Given the complexity of the soil N cycle and its interactions with other ecosystems — coordinated planning by agronomists, soil scientists, horticulturists, silviculturists, atmospheric chemists, engineers, hydrogeologists, microbiologists, resource econo-mists, and others will be needed to meet this challenge. In the following sections we will focus on the *process* of efficient N management — primarily for agricultural crops — although some references to forest soils, horticultural operations, turf, and land reclamation will be used to illustrate key differences in N management required for these situations.

4.4.1 General Principles of Efficient N Use

Approaches to assess the efficiency of N use by agricultural crops normally include both agronomic and environmental components. We seek to maximize agronomic efficiency by producing greater yields with less N and environmental efficiency by minimizing the escape of added N from the soil to an ecosystem sensitive to N pollution. Quantitative assessment mechanisms are therefore essential if we are to evaluate the success of existing and proposed management programs. Crop *N use efficiency* (NUE) from an agronomic perspective (production and eco-nomics) has traditionally been expressed in terms of *yield efficiency* (YE) or *N recovery efficiency* (NRE), defined as:

$$YE = \frac{[\text{Crop yield}]_{+\text{N Source}} - [\text{Crop yield}]_{\text{Soil alone}}}{\text{Total N added by N source}} \qquad (4\text{-}18)$$

$$NRE = \frac{[\text{Crop N uptake}]_{+\text{N Source}} - [\text{Crop N uptake}]_{\text{Soil alone}}}{\text{Total N added by N source}} \qquad (4\text{-}19)$$

Yield efficiency considers only the relationship between N fertilization and crop production and is primarily an economic assessment. Nitrogen recovery efficiency provides an estimate of applied N that, because it was not taken up by the crop, has been redistributed to some other component of the N cycle. Nitrogen recovery efficiency thus addresses both economic and environmental concerns; and if com-bined with other data, such as changes in soil NO_3^- concentrations in the crop rooting zone, can help identify the fate of the unrecovered N. Several key points should be kept in mind when evaluating NRE values for differing crops and management practices. First, it is important to remember that unrecovered N is not necessarily lost from the soil-plant system; it may be temporarily immobilized as organic N, remain in the rooting zone as NH_4^+ or NO_3^-, or even be recycled through irrigation waters. In these situations, the unrecovered N from one crop may be recovered through uptake by a subsequent crop. Second, as shown in Figure 4-13, the curvilinear nature of crop response (yield and N uptake) to applied N means that estimates of YE and NRE will vary with N rate. While, generally speaking, YE and NRE decrease and the potential for N loss to other ecosystems increases as N rate increases, the percentage of applied N that is not recovered by the crop definitely depends on the amount of N used. Finally, as most studies base NRE on the difference in N uptake between a fertilized crop and a nonfertilized control, it should be noted that applying N to a crop has been shown to affect the uptake of native soil N, both by altering soil N transformations and enhancing the ability of the crop rooting system to recover N.

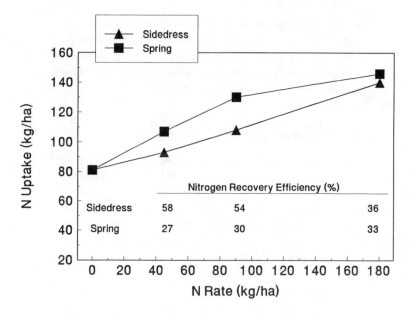

Figure 4-13 Nitrogen uptake, and N recovery efficiency (NRE) for corn. Fertilizer N was either applied in a single spring application or in a split application using starter and sidedress fertilizers. (Adapted from Bock, 1984.)

The influences and interactions of soil type, crop management (tillage, irrigation, rotation, manure use), and method and timing of N application on NRE have been assessed for many N sources and agronomic crops; and representative examples are given in Table 4-8. The purpose of these studies has normally been to identify the magnitude of N loss in different environmental settings and under changing management practices. More sophisticated research has attempted to determine the specific fate of applied N, as shown in Figure 4-14, often relying on the use of N-15 labeled fertilizers or crop residues to trace the movement of N.

The environmental efficiency of N use, while clearly important, is much more difficult to quantify, primarily because of the difficulty in establishing direct linkages between agricultural management practices and polluted resources. Most studies have used the NRE approach and the general assumption that if a management system has low NRE values (high amounts of unrecovered N) and the potential to adversely impact the environment, the system requires improvement. If some direct measurement of N pollution in the area dominated by the system is also available (e.g., high NO_3^- concentrations in wells), then the pressure to modify the system is particularly great. An example of such a scenario would be the use of poultry manure in crop production on the Delmarva peninsula (Delaware-Maryland-Virginia), where data from well surveys indicated widespread contamination of shallow groundwater wells with NO_3^-. Soils in the region are coarse-textured and well-drained, rainfall is plentiful (~100 cm/year), and use of overhead irrigation is increasing. The area contains one of the most highly concentrated poultry industries in the world, and has an agriculture dominated by crops with high N requirements (corn, wheat, barley). Laboratory studies showed that poultry manure N was rapidly converted to NO_3^- in

Table 4-8 Representative Values for N Use Efficiency of Corn

Nitrogen rate (kg/ha)	Treatments studied and N use efficiency (%)	
Delaware	Poultry manure	NH_4NO_3
84[a] (107[b])	50	62
168 (214)	33	57
252 (321)	34	50
Maryland	Minimum tillage	Plow tillage
90[a]	53	67
180	53	52
270	53	38
Vermont	Dairy manure + NH_4NO_3	NH_4NO_3 alone
0[a] (243[b])	41	—
56 (243)	35	97
112 (243)	18	93
168 (243)	16	68
Wisconsin	Sewage sludge — Site 1	Sewage sludge — Site 2
340[b]	17	21
680	19	17
1360	12	10
2720	9	6

Source: Sims, 1987 (Delaware); Meisinger et al., 1985 (Maryland); Jokela, 1992 (Vermont); and Kelling et al., 1977 (Wisconsin).

Note: Nitrogen use efficiency defined as: $\dfrac{(\text{N uptake})_{TN} - (\text{N uptake})_{noN}}{\text{Total N added}}$

[a] Rates of NH_4NO_3 added.
[b] Total amount of N added in manure or sludge.

most soils, field studies with irrigated corn found NRE values of 20–50%, and monitoring studies of wells in manured fields found NO_3^- concentrations in excess of current water quality standards (10 mg NO_3^- N/L). Although it was recognized that other sources, primarily N fertilizers and rural septic systems were contributing to groundwater NO_3^-, the potential for the poultry-based agriculture of the area to contaminate surface- and groundwaters has resulted in intensive efforts by advisory and regulatory agencies to improve crop N management, in general, and manure management, in particular.

In summary, improving the efficiency of N management will require an integrated approach, often regional in nature. Traditional approaches have combined the development and implementation of research-based *best management practices* (BMPs) for soil conservation and nutrient management with long-term basic research to alter fundamental aspects of the N cycle (Table 4-9). Recently, intense environmental pressures to control N pollution have raised the issue of mandatory, legislated controls on fertilizer N use. Many states with localized animal production facilities are reexamining the scale of management required to improve N use efficiency, recognizing that practices which optimize N recovery at the individual farm field level may be of little use if the nutrients generated on the farm exceed its capacity to use them. The following sections provide specific examples of techniques used to improve N use efficiency in agriculture and nonagricultural systems, primarily at the field level. As N pollution reflects the combined effect of many farm fields as well as urban areas, municipal and industrial waste disposal sites, land reclamation projects, and rural septic systems, it is important to view these practices from a broader perspective and to integrate them into larger scale N management programs.

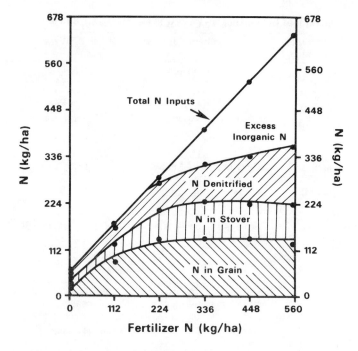

Figure 4-14 Influence of N rate on the fate of fertilizer N. (Adapted from Boswell et al., 1985. With permission.)

4.4.2 Nitrogen Management in Agriculture

Current approaches to N management in agricultural cropping systems normally begin with an assessment of the crop N requirement at a realistic yield goal. Fertilizer or organic waste management practices, based on local soil and climatic conditions, are then relied on to minimize the extra amounts of N added in anticipation of N losses (system inefficiency) and to attain optimum yields. The amount of supplemental N needed by a crop is related to the potential for excess N in the soil by the following equation, commonly referred to as a *soil nitrogen budget:*

$$N_f = \left[N_{up} \right] - \left[N_{som} + N_{na} \right] + N_{ex} \qquad (4\text{-}20)$$

where

N_f = Amount of N needed from fertilizer, manure, etc.
N_{up} = Crop N requirement at desired yield
N_{som} = N added from mineralization of soil organic matter, crop residues, previous applications of organic wastes, etc.
N_{na} = Natural additions of N (rainfall, irrigation, dry deposition)
N_{ex} = Excess N lost by denitrification, erosion, leaching, or volatilization; varies as a function of the efficiency of each soil-crop system

Minimizing N_{ex}, therefore, requires that we do not overestimate the crop yield potential for a particular soil, and thus both N_{up} and N_f, — the crop N requirement

Table 4-9 Summary of Best Management Practices (BMPs) for Efficient N Use

Management approach	Examples and comments
Soil, crop, and water management	
Soil and water conservation	Contour plowing, terracing, reduced tillage, improved irrigation management; all act to reduce erosion, runoff, and leaching of N
Cropping sequence and cover crops	Rotating legumes and nonlegumes to reduce need for N fertilizers; legumes and winter annual cover crops can "scavenge" residual soil N; benefits from crop rotations include economic stability, erosion control, reduced pest and disease pressure
Watershed management	Soil and water conservation and nutrient management supported by widespread educational programs, cost sharing, and guidelines or regulations on irrigation, fertilizer use
Nutrient management	
Soil, plant, and waste testing	Recent advances in soil and plant N testing (e.g., presidedress soil nitrate test, leaf chlorophyll meter) provide opportunities for more efficient use of fertilizers and manures
Application timing and method	Split applications of N, fertigation, slow-release fertilizers, injection and deep placement of volatile N fertilizers; all directed toward improving synchrony between N availability and crop N uptake pattern
Fertilizer and waste technology	Nitrification and urease inhibitors improve efficiency of fertilizer N recovery; composting and pelletizing stabilize N in organic wastes and provide materials that can be handled and applied more efficiently
Fundamental changes in agriculture	
Legislation	Regulations, not guidelines, mandate amount and timing of N application from fertilizers or organic wastes; farm scale nutrient budgets require farmers to find alternative uses for excess manure; most approaches are voluntary at present but legislation affecting nutrient management has been introduced in several states
Genetic advances	Genetic alteration to introduce biological N-fixation ability into nonlegumes (corn, wheat); increase N-fixation capacity of legumes; advances in genetic manipulation of plants make these long-term goals of basic research in N fixation more feasible
Cropping patterns	Increased use of legumes, decreased production of cereal grains; more legumes would reduce fertilizer N use, but likely affect economics of production, dietary habits of consumers; pressures to convert to low-input, "sustainable" agriculture with less reliance on fertilizers and pesticides; low-input agriculture is often labor intensive, while the availability of farm labor is already inadequate in many urban societies; loss of cropland due to urbanization, desertification; as food production is highly dependent on amount of arable land, major losses of cultivated acreage would increase pressure to obtain higher yields on remaining land, requiring higher inputs and increasing potential for nonpoint source pollution; many urban areas and areas of intense animal production already have inadequate land available to use organic wastes they produce

Source: Keeney, 1982.

and amount of fertilizer needed to attain the desired hopefully realistic yield. We should not underestimate the potential of the soil or other natural sources of N (N_{som}, N_{na}) to provide a significant percentage of N_{up}. Unfortunately, there are many examples where both types of errors have occurred, frequently resulting in groundwater contamination by NO_3^-. The issue of overapplication of N is complicated not only by overly optimistic estimates of potential yield, but also by the relatively inexpensive nature of fertilizer N; and when manures and sludges are involved, there is continuing pressure to dispose of organic wastes, regardless of the true N requirement of the crop. The relationship between unrealistic yield goals and potential for groundwater contamination is clearly shown in Figure 4-15 for irrigated corn production in Nebraska, and the potential for a serious imbalance in a regional N budget is illustrated in Table 4-10 for the poultry industry in Delaware.

In addition to overestimates of N_f, another serious problem for N use efficiency has been the failure to develop reliable tests for available soil N and N added in organic wastes so that farmers can accurately adjust fertilizer N rates to compensate for what the soil and other soil amendments provide. The lack of N testing procedures has not been due to a lack of research effort, however; and recently major advances have been made in the area of soil and plant testing for N.

An accurate soil test for N has been a long but elusive goal for soil scientists. The complex and dynamic nature of N cycling and its extreme sensitivity to often unpredictable climatic factors such as temperature and rainfall have made it difficult to use chemical extractants to estimate N availability in advance of planting, as is commonly done for other plant nutrients (e.g., P, K, Ca, Mg, Mn, Cu, and Zn). Similar problems have prevented the adoption of rapid chemical tests for available N in organic wastes. Residual tests for soil NO_3^--N have had a history of success in arid zone soils, but not in humid regions. In 1984 a significant breakthrough in soil N testing occurred that has shown the potential for markedly improving the efficiency of N use for certain agronomic crops. The "pre-sidedress soil nitrate test" (PSNT) was conceived and first evaluated by Dr. F. R. Magdoff of the University of Vermont to address the problem of overfertilization of corn with N in the northeastern U.S., particularly in fields with histories of manure and legume use. The PSNT has four basic tenets, briefly summarized as follows: (1) all fertilizer N for corn, except a small amount banded at planting, should be sidedressed when the crop is beginning its period of maximum N uptake; (2) soil and climatic conditions prior to sampling integrate the factors influencing the availability of N from the soil, from crop residues, and from previous applications of organic wastes; (3) a rapid sample turnaround (<14 day) by a testing laboratory is possible; and (4) farmers will normally not sample to a depth >30 cm. The PSNT has since been evaluated in over 300 field studies in the northeastern and midwestern U.S. and has been repeatedly shown to be successful in identifying N sufficient soils (Figure 4-16a). Some of the logistical difficulties associated with the need for a rapid sample analysis have been overcome by the development of "quick-test" kits and NO_3^--sensitive electrodes that can be used in the field. Even more encouraging are the results of a recent study with the leaf chlorophyll meter which showed that this extremely rapid, in-field

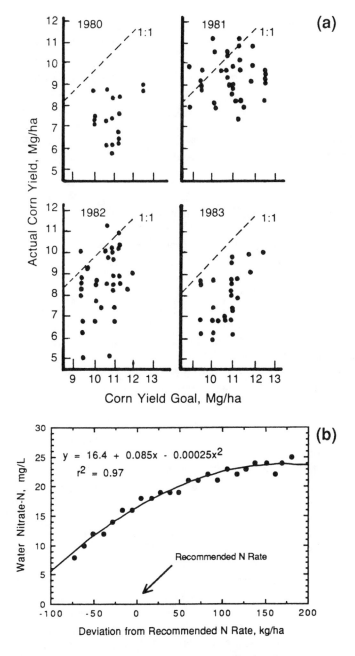

Figure 4-15 (a) Comparison of farmer yield goals for irrigated corn with actual yields obtained. Dashed line indicates 1:1 fit between yield goal and actual yield. (Adapted from Hergert, 1987. With permission.) (b) Influence of deviation from recommended N rate on groundwater nitrate concentrations in Nebraska. (Adapted from Schepers et al. 1991. With permission.)

Table 4-10   Statewide N Budget for Poultry-Based Agriculture in Delaware

Crop	Hectares	Annual N requirement (kg/Crop)	N Source	Annual amount of available N (kg/Source)
Corn	69,600	9,800,000	Poultry manure	7,850,000
Soybeans	80,600	0	Fertilizer sales	19,350,000
Wheat	24,300	2,200,000	Other wastes	Undocumented
Barley	11,000	1,000,000		
Other	32,400	3,600,000		
Total	217,900	16,600,000	Total	27,200,000+

Note: Statewide N balance: an estimated excess of 10,600,000 kg/state/yr or ~50 kg/ha/yr.

measurement of leaf "greenness" was as accurate as the PSNT in identifying N-sufficient sites (Figure 4-16b). Another new approach to assessing N sufficiency for corn is the stalk nitrate test (Figure 4-16c). This "postmortem" test uses the concentration of NO_3^- in the lower portion of the corn stalk at maturity to identify fields that received excessive N from fertilizers or manures.

The implications of these N tests are straightforward, but not simple. For most farmers a PSNT soil sample would be taken and, if necessary, additional fertilizer would be applied via sidedressing. However, studies from soils commonly amended with animal wastes have shown that often little or no sidedress N is required, even when manure was not applied in the current year. This is illustrated in Figure 4-17 where the *economically optimum N* (EON) *rates* are compared for 11 fields with and without long-term histories of manure use. EON rates are analogous to yield efficiency and are defined as the N rate where economic return on fertilizer N investment is maximized, based on assumed fertilizer costs and crop prices. The average EON rates for manured fields were 34 vs 128 kg N/ha for nonmanured sites. The greatest difficulty with the PSNT approach — apart from logistical problems associated with the rapid analytical turnaround — has been the presence of high percentages of soils that have been shown to need less, or no, manure/sludge than is generated by the farm or municipality. Simply put, these tests have shown that — particularly for animal-based agriculture — more N is produced than is needed by the farming operation, given the land available and the economics of waste handling and application. This once again illustrates the need for organic waste management at a larger scale, state or regional in scope, oriented toward redistribution of waste N to nutrient-deficient areas.

Once the need for and rate of supplemental N have been determined, either through soil and plant testing or a general knowledge of the cropping system, the next considerations are the source of N to be used and the method and timing of N application. Maximization of N efficiency requires that the best N source be applied in as timely a manner as possible, often through the use of multiple applications. As mentioned in Section 4.3, the primary consideration in selection of an N source is often economics, beginning with cost per unit of N. Other factors that influence the total cost of the N source include availability of the material (transportation costs), storage and handling (equipment and application costs), crop-specific fertilization requirements (e.g., the need for a mixed fertilizer), and any properties of the fertilizer material that may significantly affect its efficiency in the cropping system used. In

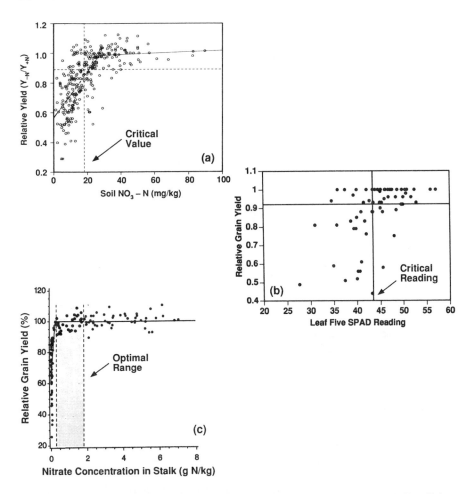

Figure 4-16 Illustration of the use of soil and plant N testing procedures to identify N sufficient soils. (a) Relative yield of corn (Y_{-N}, yield in control (0 N) treatment, divided by Y_{+N}, yield from adequately fertilized treatment) vs soil nitrate (0–30 cm) in the late spring. (Adapted from Magdoff et al., 1990. With permission.) (b) Relative yield of corn vs leaf chlorophyll meter reading taken in late spring. SPAD refers to type of meter used to make reading. (Adapted from Piekielek and Fox, 1992. With permission.) (c) Relative yield of corn vs nitrate concentration in the lower portion of the stalk at maturity. (Adapted from Binford et al., 1990. With permission.)

animal-based agriculture where manure is constantly produced, other costs arise such as the need for manure analysis, manure storage and treatment (e.g., lagoons, storage barns, composting facilities), and specialized equipment for manure application.

Selection of a N source can rarely be separated from selection of a N application method. Typical methods used to apply N fertilizers include *broadcasting, banding, injection,* and *fertigation. Broadcasting* refers to the uniform applications of fertilizer materials to the soil surface, as solid granules or pellets or as a liquid spray. The broadcast fertilizer may then remain on the surface (no-tillage crops, turf, forages) or be incorporated by a tillage operation such as disking or plowing. Broadcast

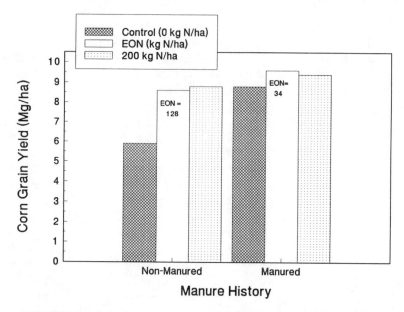

Figure 4-17 Yield response of corn to economically optimum rates of N (EON) in fields with and without a history of manure applications. Numbers in bars are average yields (in Mg/ha) of all nonmanured or manured sites at EON rate. (Adapted from Roth and Fox, 1990.)

applications made to a growing crop are referred to as *topdressing. Banding* is the placement of fertilizer N in a narrow band and can be done at planting; for row crops, banding fertilizers after crop emergence is referred to as *sidedressing.* Band applications of N at planting normally use N-P "starter" fertilizers with placement of the fertilizer approximately 5 cm below and 5 cm to the side of the seed to avoid salt injury to young seedlings. Banding at planting places an initial supply of N and P in a zone highly accessible to young root systems, and is particularly useful in no-tillage soils where cooler soil temperatures often delay initial root growth and nutrient uptake. *Sidedress* applications of N fertilizers can be done through injections, surface bands, sprays, or "dribbles." *Injection* is the placement of fertilizer in the soil through specialized subsurface application equipment and is most commonly used for anhydrous NH_3 and liquid fertilizer materials such as urea-ammonium-nitrate (UAN) solutions. Fertilizer injections are normally made at deeper depths than band placements, particularly for anhydrous NH_3 which reverts to a gas following injection and can be lost from the soil via volatilization if not placed correctly. *Fertigation* is the application of soluble N in irrigation waters and is not the same as *foliar fertilization* where fertilizers are sprayed directly on plant foliage. Only a limited number of N fertilizers are suitable for use in fertigation systems, due to the high degree of solubility and purity required to ensure complete dissolution and avoid clogging of application equipment. UAN solutions are normally preferred over anhydrous or aqua NH_3 which has a much higher volatilization potential. Fertigation is most commonly used with overhead sprinklers or drip irrigation systems.

The method of application selected for an organic N source depends primarily on its physical properties. Solid organic wastes are usually applied by large flail or spinner spreaders, although recent advances in waste processing (such as pelletizing), have increased the flexibility available to waste applicators. Liquid organic wastes either are injected or applied through wastewater irrigation systems. Incorporation of organic wastes is often mandated to avoid potential runoff of nutrient-rich solids or soluble organic materials into streams and lakes.

One of the most critical aspects of N use efficiency is the timing of N application relative to crop N uptake. Most crops, particularly annual crops, have well-defined patterns of N accumulation, as illustrated for corn in Figure 4-18. Fertilizer application techniques that deliver supplemental N in close synchrony with N uptake will, in general, be the most efficient, particularly on soils that are highly sensitive to N losses. A generalized ranking of the relative efficiency of the most common N application techniques would be fertigation > banding ~ sidedress > surface broadcast. The greatest efficiency of N application normally results when several application techniques are combined. For irrigated corn this might involve a small amount of fertilizer N applied at planting, a sidedress application providing ~30–40% of the N requirement immediately prior to the period of most rapid N uptake, and several fertigations that supply the remainder of the crop requirement (Figure 4-18a). In a similar situation using organic wastes this could involve a low (suboptimum N rate) application rate of manure shortly before planting, followed by use of the PSNT to identify the sidedress N requirement (Figure 4-18b). There are many situations, however, where serious limitations to improved timing of application of an N source exist. One of the greatest of these limitations is the amount of time required for delayed or multiple applications of N during the part of the growing season when many other operations must be performed in a timely manner, including tillage, planting, herbicide application, harvest of other crops, and installation of irrigation systems. As an example, for many farmers applications of fertilizer or manure N during fall and winter months would be preferred, because of the greater amount of free time available during this part of the year. However, the low efficiency of N recovery during these periods because of leaching and denitrification losses that can occur prior to crop growth in the spring and summer make this undesirable both economically and environmentally. Manure applications are often preferred during the fall or winter because the dry or frozen nature of the soil at this time can reduce the amount of compaction and erosion caused by heavy equipment traffic in fields relative to wet, spring conditions.

The basic reason that applications far in advance of crop N uptake are undesirable is the rapid rate of nitrification in most soils. Thus the NH_4^+ applied in most fertilizers or manures is normally found in the soil as NO_3^- within a short period of time and is subject to leaching or denitrification losses prior to the initiation of crop uptake. One means to improve N use efficiency is to delay the process of nitrification by the use of chemical inhibitors applied in conjunction with N fertilizers or organic wastes, as shown in Figure 4-19. Chemicals such as nitrapyrin and thiosulfate have been shown to inhibit the activity of nitrifying organisms, keeping applied N in the less leachable, ammoniacal form and increasing the possibility of crop N uptake. Slow-release N

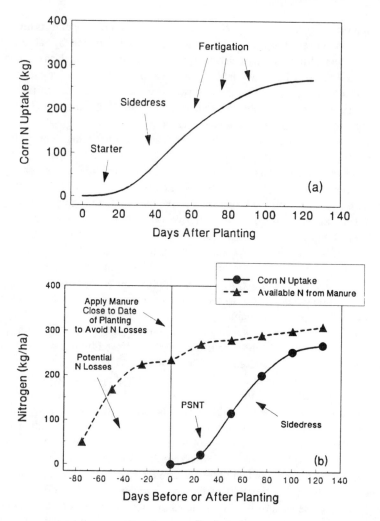

Figure 4-18 (a) Generalized representation of N uptake pattern for corn, illustrating an efficient combination of N fertilizer techniques. (b) N uptake pattern for corn combined with typical rate of N release from poultry manure. Illustrates the potential for N losses from poorly timed manure applications and the timing of the pre-sidedress soil nitrate test, relative to sidedress N application.

sources achieve the same goal by physically sealing the NH_4^+ within a resin, wax, or coating of sulfur; this delays the dissolution of the fertilizer granule and, by doing so, the process of nitrification.

Nitrogen use efficiency can be improved by other means as well, although practical manipulations of N loss mechanisms can be difficult and expensive, and may increase one form of loss while reducing another. The use of conservation tillage practices can be expected to reduce erosion and runoff losses of N. Reducing water movement off a field, however, will likely increase infiltration and thus NO_3^- leaching and denitrification. Surface applications of wastes may also reduce soil-waste contact

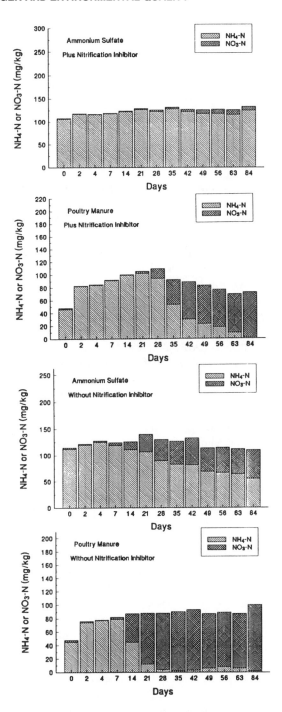

Figure 4-19 Illustration of the influence of a nitrification inhibitor (thiosulfate) on the rate of nitrification of an inorganic N fertilizer (ammonium sulfate) and an organic N fertilizer (poultry manure). (Adapted from Sallade and Sims, 1992.)

and accelerate waste drying, enhancing NH_3 volatilization but decreasing the rate of N mineralization. Other conservation practices that have the potential to reduce N losses include more efficient irrigation practices (e.g., drip versus flood), use of multiple cropping or winter cover crops to trap residual N from wastes, and controlled drainage systems or artificial wetlands to enhance denitrification in field border areas.

4.4.3 Nitrogen Management in Situations Other Than Production Agriculture

Nitrogen fertilizers and organic N sources are widely used in situations that differ greatly from production agriculture, such as land reclamation, road construction, urban and commercial horticulture, and forestry. Each of these end uses has its own particular constraints with regard to N use efficiency. Some are related to differences in the nature of the N cycle in disturbed or heavily amended soils; and others to the economics, politics, and logistics of soil-plant management. Reclamation and construction projects normally aim to produce a perennial, low maintenance ground cover that will only require the use of N fertilizers initially, or at most in small, infrequent applications. Horticultural situations such as greenhouses, nurseries, turf, and ornamental plantings are usually much more intensive in nature; and because of the high cash value of the plants involved, these rely heavily on fertilizer, often with little concern for the economic or environmental efficiency of N use. Forest fertilization with N is most common in the commercial forestry industry and represents an intermediate situation between low input reclamation/construction projects and high input commercial horticulture.

Organic N sources are commonly used in both of these nonagricultural settings for a number of reasons. The organic matter provided by sludges, composts, and occasionally animal manures is often needed in reclamation projects to improve the deteriorated physical properties of highly disturbed soils. Greenhouses use high organic matter, "soilless" media almost exclusively to provide a light, well-aerated growth medium with high moisture-holding capacity. Both situations are good end uses for municipal organic wastes, such as sludge composts, because they predominantly use nonfood chain plants, thus concerns about dietary accumulations of potentially toxic waste constituents (e.g., heavy metals) are reduced.

Management of soil, fertilizer, and organic N in land reclamation and forestry situations begins with an understanding of the differences that can occur in N cycling, relative to cultivated soils. These differences can be fundamental or practical in nature. At the most basic level, the microorganisms responsible for key N transformations such as mineralization, nitrification, and denitrification may be less efficient, inactive, or nonexistent in soils that have extreme physical or chemical properties (such as highly acidic mine spoils or pine forests). From a practical standpoint, amending these soils with fertilizer, organic N, or even small intermittent doses of N as acid rain can alter the activities of these soil microbes and change the nature of N cycling (shown in Figure 4-20 for nitrification in sludge-amended forests and Table 4-11 for nitrification and denitrification in acid mine spoils).

From a horticultural perspective, there has been increasing interest in the environmental implications of N management for greenhouses, nurseries, and turf. The

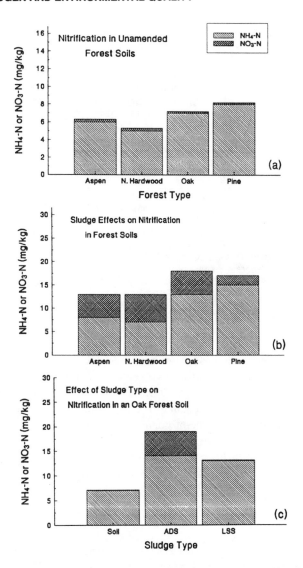

Figure 4-20 Nitrification in forest soils. (a) Effect of forest type on nitrification in unamended forest soils. (b) Amending forest soils with sludge increases nitrification. (c) Sludge type affects extent of nitrification in an oak forest (ADS = anaerobically digested sludge, LSS = lime stabilized sludge). (Adapted from Burton et al., 1990.)

"soilless" media used in many greenhouses and nurseries is not relied on to provide much of the plant N requirement, serving mainly as a physical growth medium. Growers instead commonly use fertigation with high N concentration solutions combined with intensive watering to remove salts from containers. This has raised questions about the impact of drainage waters from these facilities on surface- and groundwater quality. More efficient fertilization programs are available and are being adopted in response to these environmental concerns. Examples include the use of slow-release fertilizers, "ebb-and-flow" fertigation systems where the nutrient solutions are recycled and reused, and on-site wastewater treatment systems. Similarly,

Table 4-11 Effect of Amending Highly Acidic Mine Spoils on Nitrification, Denitrification, and Populations of Nitrate-Reducing Bacteria

Mine spoil	pH	N Mineralization potential[a]		Denitrification parameters[b]	
		NH$_4$-N[c]	NO$_3$-N[c]	DEA[d]	MPN[e]
Bald Knob					
Unamended	3.8	21	3	5	3
Amended	6.8	−12	38	11	180
Osage					
Unamended	2.7	33	8	11	6
Amended	5.2	−8	21	68	980

Source: Shirey and Sexstone, 1989.
Note: Bald Knob site amended with lime and fertilizer, Osage site with coal fly ash.
[a] Rate of change in NH$_4$-N and NO$_3$-N in mine spoil materials during a 30-d laboratory incubation study.
[b] Rate of denitrification of added NO$_3$ from mine spoils as determined in laboratory study of denitrifying enzyme activity (DEA); MPN is estimate of most probable number of nitrate reducers in mine spoil material based on five-tube assay.
[c] µg/kg/hr.
[d] µg/kg/hr.
[e] MPN/g.

improved N management programs for turf have been developed and promoted more intensively to homeowners and those involved in industrial or recreational turf management (e.g., golf courses, parks, athletic fields). Avoiding N losses by leaching or runoff is only part of the concern with turf. Overfertilization with N increases the frequency of cutting and the volume of grass clippings that must be disposed of in landfills. Because many municipalities are no longer accepting yard wastes, the pressure to avoid the use of excess N and to recycle N "on-site" through proper cutting schedules, mulching, or composting has increased.

REFERENCES

Alberts, E. E., Neibling, W. H., and Moldenhauer, W. C., Transport of sediment nitrogen and phosphorus through cornstalk residue strips, *Soil Sci. Soc. Am. J.*, 45, 1177, 1981.

Berry, J. T. and Hargett, N. L., 1986 Fertilizer Summary Data, National Fertilizer and Environmental Research Center, Tennessee Valley Authority, Muscle Shoals, AL, 1986, 132 pp.

Binford, G. D., Blackmer, A. M., and El-Hout, N. M., Tissue test for excess nitrogen during corn production, *Agron. J.*, 82, 124, 1990.

Bock, B. R., Efficient use of nitrogen in cropping systems, in Nitrogen in Crop Production, Hauck, R. D., Ed., American Society of Agronomy, Madison, WI, 1984, 273.

Boswell, F. C., Meisinger, J. J., and Case, N. L., Production, marketing, and use of nitrogen fertilizers, in *Fertilizer Technology and Use*, Engelstead, O. P., Ed., American Society of Agronomy, Madison, WI, 1985, 229.

Burton, A. J., Hart, J. B., and Urie, D. H., Nitrification in sludge-amended Michigan forest soils, *J. Environ. Qual.*, 19, 609, 1990.

Chae, Y. M. and Tabatabai, M. A., Mineralization of nitrogen in soils amended with organic wastes, *J. Environ. Qual.*, 15, 193, 1986.

Deans, J. R., Molina, J. A. E., and Clapp, C. E., Models for predicting potentially mineralizable nitrogen and decomposition rate constants, *Soil Sci. Soc. Am. J.,* 50, 323, 1986.

Donahue, R. L., Miller, R. W., and Shickluna, J. C., *Soils: An Introduction to Soils and Plant Growth,* 5th ed., Prentice-Hall, Englewood Cliffs, NJ, 1983.

Donovan, W. C. and Logan, T. J., Factors affecting ammonia volatilization from sewage sludge applied to soil in a laboratory study, *J. Environ. Qual.,* 12, 584, 1983.

Ellert, B. H. and Bettany, J. R., Comparison of kinetic models for describing net sulfur and nitrogen mineralization, *Soil Sci. Soc. Am. J.,* 52, 1692, 1988.

Epstein, E., Keane, D. B., Meisinger, J. J., and Legg, J. O., Mineralization of nitrogen from sewage sludge and sludge compost, *J. Environ. Qual.,* 7, 217, 1978.

Fox, R. H. and Hoffman, L. D., The effect of N fertilizer source on grain yield, N uptake, soil pH, and lime requirement in no-till corn, *Agron. J.,* 73, 891, 1981.

Gartley, K. L. and Sims, J. T., Ammonia volatilization from poultry manure-amended soils, *Biol. Fert. Soils,* 16, 5–10, 1993.

Hadas, A., Bar-Yosef, B., Davidov, S., and Sofer, M., Effect of pelleting, temperature, and soil type on mineral N release from poultry and dairy manures, *Soil Sci. Soc. Am. J.,* 47, 1129, 1983.

Hamilton, P. A. and Shedlock, R. J., Are fertilizers and pesticides in the ground water? A case study of the Delmarva peninsula, U.S. Geological Survey Circular 1080, Denver, CO, 1992.

Hergert, G. W., Status of residual nitrate-nitrogen soil tests in the United States of America, in *Soil Testing: Sampling, Correlation, Calibration and Interpretation,* Brown, J. R., Ed., American Society of Agronomy, Madison, WI, 1987, 73.

Igo, E. C., Sims, J. T., and Malone, G. W., Advantages and disadvantages of manure analysis for nutrient management purposes, *Agron. Abstr.,* 154, 1991.

Jokela, W. E., Nitrogen fertilizer and dairy manure effects on corn yield and soil nitrate, *Soil Sci. Soc. Am. J.,* 56, 148, 1992.

Keeney, D. R., Nitrogen management for maximum efficiency and minimum pollution, in *Nitrogen in Agricultural Soils,* Stevenson, F. J., Ed., American Society of Agronomy, Madison, WI, 1982, 605.

Kelling, K. A., Walsh, L. M., Keeney, D. R., Ryan, J. A., and Peterson, A. E., A field study of the agricultural use of sewage sludge. II. Effect on soil N and P, *J. Environ. Qual.,* 6, 345, 1977.

Magdoff, F. R., Jokela, W. E., Fox, R. H., and Griffin, G. F., A soil test for nitrogen availability in the Northeastern United States, *Commun. Soil Sci. Plant Anal.,* 21, 1103, 1990.

McInnes, K. J., Ferguson, R. B., Kissel, D. E., and Kanemasu, E. T., Field measurements of ammonia loss from surface applications of urea solution to bare soil, *Agron. J.,* 78, 192, 1986.

Meisinger, J. J., Bandel, V. A., Stanford, G., and Legg, J. O., Nitrogen utilization of corn under minimal tillage and moldboard plow tillage. I. Four year results using labeled N fertilizer on an Atlantic coastal plain soil, *Agron. J.,* 77, 602, 1985.

Nielsen, E. G. and Lee, L. K., The magnitude and costs of groundwater contamination from agricultural chemicals: a national perspective, Agric. Econ. Rep. No. 576, U.S. Department of Agriculture, Washington, D.C., 1987, 38 pp.

Piekielek, W. P. and Fox, R. H., Use of a chlorophyll meter to predict sidedress nitrogen requirements for maize, *Agron. J.,* 84, 59, 1992.

Rice, C. W., Sierzega, P. E., Tiedje, J. M., and Jacobs, L. W., Stimulated denitrification in the microenvironment of a biodegradable organic waste injected into soil, *Soil Sci. Soc. Am. J.,* 52, 102, 1988.

Roth, G. W. and Fox, R. H., Soil nitrate accumulations following nitrogen-fertilized corn in Pennsylvania, *J. Environ. Qual.,* 19, 243, 1990.

Sallade, Y. E. and Sims, J. T., Evaluation of thiosulfate as a nitrification inhibitor for manures and fertilizers, *Plant and Soil,* 147, 283, 1992.

Schepers, J. S., Moravek, M. G., Alberts, E. E., and Frank, K. D., Maize production impacts on groundwater quality, *J. Environ. Qual.,* 20, 12, 1991.

Shirey, J. J. and Sexstone, A. J., Denitrification and nitrate reducing bacterial populations in abandoned and reclaimed mine soils, *FEMS Microb. Ecol.,* 62, 59, 1989.

Sims, J. T., Agronomic evaluation of poultry manure as a nitrogen source for conventional and no-tillage corn, *Agron. J.,* 79, 563, 1987.

Sims, J. T., Murphy, D. W., and Handwerker, T. S., Composting of poultry wastes: implications for dead poultry disposal and manure management, *J. Sust. Agric.,* 2(4), 67–82, 1992.

Stanford, G., Frere, M. H., and Schwaninger, D. H., Temperature coefficient of nitrogen mineralization in soils, *Soil Sci.,* 115, 321, 1973.

Stanford, G. and Epstein, E., Nitrogen mineralization-water relations in soils, *Soil Sci. Soc. Am. Proc.,* 38, 103, 1974.

Touchton, J. T. and Hargrove, W. L., Nitrogen sources and methods of application for no-tillage corn production, *Agron. J.,* 74, 823, 1982.

Vigil, M. F. and Kissel, D. E., Equations for estimating the amount of nitrogen mineralized from crop residues, *Soil Sci. Soc. Am. J.,* 55, 757, 1991.

SUPPLEMENTARY READING

Dahnke, W. C. and Johnson, G. V., Testing soils for available nitrogen, in *Soil Testing and Plant Analysis,* Westerman, R. L., Ed., Soil Science Society of America, Madison, WI, 1990, 127.

Fedkiw, J., Nitrate Occurrence in U.S. Waters (and Related Questions): A Reference Summary of Published Sources from an Agricultural Perspective, U.S. Department of Agriculture, Washington, D.C., 1991.

Follett, R. F., Keeney, D. R., and Cruse, R. M., *Managing Nitrogen for Groundwater Quality and Farm Profitability,* American Society of Agronomy, Madison, WI, 1991.

Legg, J. O. and Meisinger, J. J., Soil nitrogen budgets, in *Nitrogen in Agricultural Soils,* Stevenson, F. J., Ed., Agron. Monogr. 22, American Society of Agronomy, Madison, WI, 1982, 503.

Power, J. F. and Schepers, J. S., Nitrate contamination of groundwater in North America, *Agric. Ecosystems Environ.,* 226, 165, 1989.

Sommers, L. E. and Giordano, P. M., Use of nitrogen from agricultural, industrial and municipal wastes, in *Nitrogen in Crop Production,* Hauck, R. D., Ed., American Society of Agronomy, Madison, WI, 1984, 207.

5 SOIL PHOSPHORUS AND ENVIRONMENTAL QUALITY

5.1 PHOSPHORUS AND THE ENVIRONMENT

Phosphorus (P) is essential to all forms of life on earth and has no known toxic effects. Environmental concerns associated with P center on its stimulation of biological productivity in aquatic ecosystems. In most surface waters such as lakes, ponds, and bays the growth of algae or aquatic plants is limited by inadequate levels of P. Large inputs of P from urban wastewater systems, surface runoff, or subsurface groundwater flow can remove this limitation and increase the aquatic biomass to ecologically undesirable levels. *Eutrophication,* defined as *an increase in the fertility status of natural waters that causes accelerated growth of algae or water plants,* can then result. The negative effects associated with eutrophication of surface waters are important from both economic and environmental perspectives. As nutrient inputs to surface waters gradually increase, the trophic state of the water body passes through four stages of eutrophication, *oligotrophic, mesotrophic, eutrophic,* and *hypereutrophic.* At each stage, progressive changes in the ecology of the water bodies occur, affecting, usually in a negative manner, their economic and recreational uses. The properties and environmental problems of eutrophic waters are summarized in Chapter 2 (Tables 2-3 and 2-4).

Management of P to avoid eutrophication requires an understanding of several key aspects of P in the environment:

- What are the characteristics of the water body that may control the potential for eutrophication to occur?
- What are the natural inputs of P into the water body?
- What inputs of P are occurring from agricultural, urban, and industrial sources?
- How does the soil P cycle affect the availability of P for transport to the water body?
- What are the dominant transport mechanisms operational between the sources of P and the water body?
- What management practices can be used to reduce P loading to the water body from all sources, urban and rural?
- How can an integrated, economically feasible approach be developed to reduce P loading to the water body?

5.1.1. Eutrophication: The Role of Phosphorus

As the sole environmental impact of P is its role in eutrophication, a clear understanding of this process is essential to the development of sound strategies for P management. In oligotrophic lakes, soluble and sediment-bound nutrient (N, P) concentrations are low and limit the growth of algae and other water plants. These lakes are normally deeper and have adequate oxygen levels even in the summer when photosynthesis and temperatures favor maximum plant growth. They also have a high degree of species diversity for plants and animals. However, oligotrophic lakes, because of their low nutrient concentrations, have low biological productivity; and if used for the economic production of fish, often require continuous fertilization. Alterations in the ecosystem (i.e., construction of wastewater treatment plants, conversion of land from forested to agricultural or urban use) can increase inputs of nutrients to oligotrophic lakes, especially N and P, stimulating the growth of algae and other water plants and initiating the eutrophication process. As illustrated in Figure 5-1, converting land that is 90% forested to 90% agricultural, could result in a threefold increase in total P loss in runoff.

Once eutrophic conditions are established, algal blooms and other ecologically damaging effects can occur, including low dissolved oxygen levels, excessive aquatic weed growth, increased sedimentation, and greater turbidity. Decreased oxygenation is the primary negative effect of eutrophication because low dissolved oxygen levels seriously limit the growth and diversity of aquatic biota and, under extreme conditions, cause fish kills. The increased biomass resulting from eutrophication causes the depletion of oxygen, especially during the microbial decomposition of plant and algal residues. Under the more turbid conditions common to eutrophic lakes, light penetration into lower depths of the water body is decreased, resulting in reduced growth of subsurface plants and benthic (bottom-living) organisms. In addition to ecological damage, eutrophication can increase the economic costs of maintaining surface waters for recreational and navigational purposes. Surface scums of algae, foul odors, insect problems, impeded water flow and boating due to aquatic weeds, shallower lakes that must be dredged to remove sediment, and disappearance of desirable fish communities are among the most commonly reported undesirable effects of eutrophication. These environmental impacts have resulted in major efforts to develop management strategies to reduce nutrient inputs to surface waters, and, where possible, to reclaim eutrophic lakes or ponds.

Controlling eutrophication requires that both N and P enrichment of surface waters be minimized, although in most freshwaters biological productivity is limited by P, not N. Recent evidence from estuarine systems, such as the Chesapeake Bay, suggests that the limiting nutrient in tidal or saline water bodies may be N. The much lower solubility and hence bioavailability of P, relative to N, in most natural systems (terrestrial and aquatic) explains why freshwaters are normally limited by P. The eutrophication threshold for most P-limited aquatic systems is ~30 µg P/L (30 ppb). Water bodies with naturally low P concentrations will, therefore, be highly sensitive to external inputs of P from sources such as agricultural runoff, domestic sewage treatment systems, urban storm water discharge, and industrial wastewaters.

Strategies to reduce P loading must consider the nature of the P source as well as the sensitivity of the water body to eutrophication. Even natural ecosystems, such as

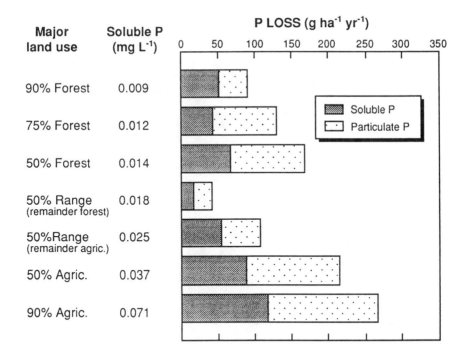

Figure 5-1 Effect of land use on the loss of soluble and particulate P from soils. (Adapted from Sharpley and Halvorson, in press.)

forests or grasslands, normally considered to be low in P fertility, may provide enough P from runoff and subsurface flow to alter the nutrient balance of a water body on the threshold of eutrophication. Controlling P inputs from these areas is generally not feasible, with the possible exception of commercial forestry operations where erosion control practices can reduce sediment loads to surface waters. In contrast, P discharge from a concentrated point source (e.g., sewage treatment plant) may be relatively easy to identify and control, but costly to correct due to the expenses associated with technological improvements to municipal or industrial wastewater treatment plants. Conversely, reducing P loading to surface waters from nonpoint sources — such as runoff from an urban or agricultural watershed — is often limited not by technological costs, but by the difficulties associated with implementing improved nutrient management practices over a large area with diverse agricultural (or industrial) enterprises. The relative magnitudes of these P inputs and the complexities involved in reducing P (or N) loading to lakes and other water bodies are illustrated in Figure 5-2 which provides a broad view of the impact of civilization on a watershed P "budget" (P outputs — P inputs). In the case of the Lake Ringsjon watershed, the budget is decidedly negative: P inputs from all sources including municipal and rural sewage, manures, fertilizers, and runoff from urban and forested areas total ~512,000 kg/yr; P outputs (crops, groundwater) are ~150,000 kg/yr. Although all these P inputs do not directly enter the lake, it seems clear that in the long-term nutrient enrichment of this lake is likely due to the magnitude of the urban and agricultural activities in the watershed.

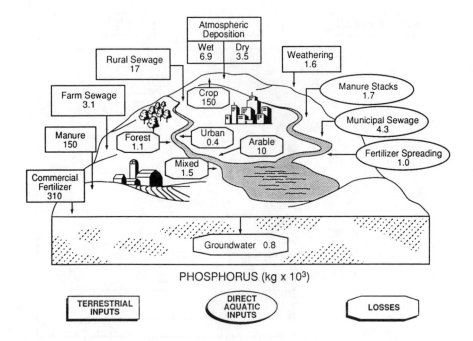

Figure 5-2 A phosphorus budget for agricultural and urban areas of the Lake Ringsjon (Sweden) watershed. (Adapted from Ryding et al., 1990.)

5.1.2 Environmental Impacts of Soil Phosphorus

Although many sources of P have the potential to induce eutrophication in surface waters, the focus of this chapter is the management of P in agricultural soils to avoid water quality problems. Nonpoint source pollution of P from urban soils will also be considered, primarily to illustrate the different management options needed for areas where erosion, storm water runoff, and intensive horticultural operations are the major sources of P.

The total quantity of P in a lake or other surface water body will be controlled by the balance between the inputs from the external sources of P described above, and the outputs as water drains from the lake via rivers, streams, or other watercourses. A net increase in P will increase the likelihood of eutrophication; however, the cycling of P between soluble, organic, and sediment-bound forms within the lake will regulate the bioavailability of P and thus the extent of eutrophication that occurs. Therefore, insofar as soil P is concerned, its importance in eutrophication will be regulated by the chemical, biological, and physical reactions that control P solubility; and the transport processes that move all forms of P to and within water bodies. An overview of these complex phenomena is given in Figure 5-3 which summarizes the movement of P in soils, the major transport mechanisms to a water body, and the major transformations that P undergoes in an aquatic system. In brief, most P in agricultural (or urban) soils is found either as insoluble precipitates of Ca, Fe, and/or Al or as a constituent of a wide range of organic compounds. Water moving across or through soils removes both soluble P and sediments enriched with P, usually the

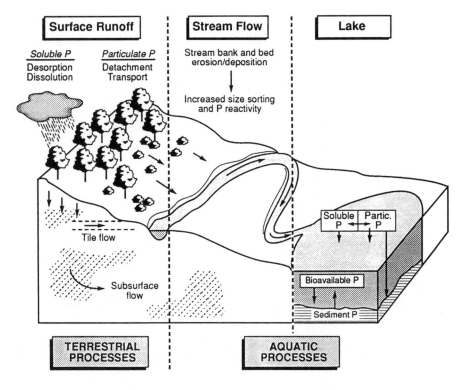

Figure 5-3 Phosphorus transport and fate in terrestrial and aquatic ecosystems. (Adapted from Sharpley and Halvorson, in press.)

lighter, fine-sized particles such as clays and some organic matter initiating the transport process. The soluble or particulate P then either can enter a flowing water body where it can be deposited as sediment or can be carried directly into a lake or pond. Phosphorus can also leach downward in the soil, perhaps to a tile drainage system or to groundwater, where subsurface transport can then discharge the P into a stream or lake. Once in the lake, the soluble P is immediately available for uptake by aquatic organisms such as algae, while much of the particulate P is removed from the lake by deposition as sediment. It is important to note, however, that P bound to sediments in streams or lakes may become available for biological uptake at a later date. This can occur when P uptake by aquatic plants depletes soluble P or when climatic conditions cause lake turbulence and a resuspension of sediments. Clearly, an in-depth understanding of the chemistry, biology, and physics of P in the soil and aquatic environment will be needed to design management practices that will reduce P impacts on water quality.

5.2 THE SOIL PHOSPHORUS CYCLE

Adequate P levels in soils are essential for the production of agricultural crops and plants grown for aesthetic purposes in urban areas. Phosphorus fertilization is thus

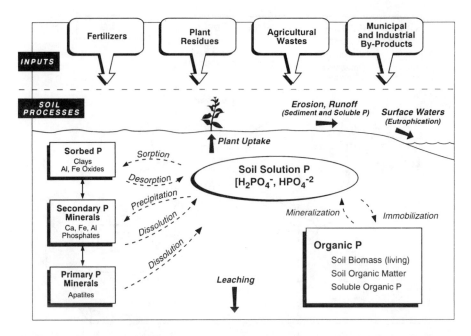

Figure 5-4   The soil phosphorus cycle. An overview of the physical, chemical, and microbiological processes controlling the availability of P to plants and P transport in runoff
or leaching waters. (Adapted from Gachon, 1969.)

a vital component of modern agriculture. Phosphorus, found in all terrestrial environments, primarily originates from the weathering of soil minerals and other more
stable geologic materials. As P is solubilized in soils by the chemical and physical
processes of weathering, it is accumulated by plants and animals, reverts to stable
forms in the landscape, or is eroded from soils and deposited as sediments in
freshwaters or oceans. Urban areas or large animal production operations convert
biologically accumulated P (food or feed) into human or animal wastes (sewage,
manures) and recycle this P into the ecosphere with varying degrees of efficiency.
Soil factors that control the rate of conversion of P between the inorganic and organic
forms regulate the short- and long-term fates of P in the environment. The soil P
cycle, illustrated in Figure 5-4, consists of many complex chemical and microbiological reactions. The overall goal of sound agronomic and environmental management programs for soil P is to maximize plant growth, while minimizing losses of P
to surface waters. It is important, therefore, to understand the role of each reaction
in controlling the availability of soil P for plant uptake or loss in erosion, surface
runoff, leaching, and drainage waters.

5.2.1 Inorganic Soil Phosphorus

Total P levels in soils normally ranges from 50 to 1500 mg/kg, with 50–70%
found in the inorganic form. The principal minerals that weather and release P into
more soluble forms differ somewhat among soils, primarily as a function of time and
soil development, as shown in Figure 5-5a. In soils that are unweathered or moderately weathered, the dominant minerals are the *apatites,* calcium phosphates with the

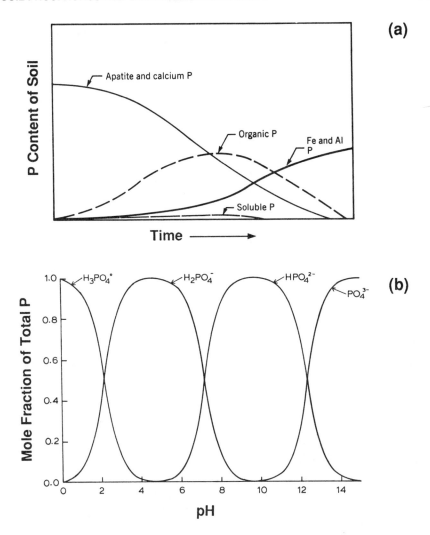

Figure 5-5 Changes in the form of soil P, as affected by: (a) time and soil development for total P; (Adapted from Foth and Ellis, 1988. With permission.) (b) soil pH for soluble P.

general chemical formula $Ca_{10}(PO_4)_6X_2$, where X represents anions such as F^-, Cl^-, OH^-, or CO_3^{2-}. Although over 200 forms of mineral P are known to occur in nature, fluorapatites are the most common form both as a mineable ore and as a component of most arable soils. In areas of intense weathering, Ca and other basic minerals are eventually leached from the soil, pH of the soil solution decreases, and iron (Fe) and aluminum (Al) solubilization occurs. Precipitates of Fe, Al, and P then form and become the major mineral sources of P in highly weathered soils. Noncrystalline oxides of Fe and Al are also common in these soils and act as "sinks" for P, through a variety of chemical reactions collectively referred to as *phosphorus fixation*.

As P is dissolved or desorbed from soil solids, it enters the soil solution, where its ionic form is largely controlled by soil pH. In the pH ranges commonly found in agricultural soils (pH 4.0–9.0), P will be found as either a monovalent ($H_2PO_4^-$) or a

Table 5-1   Phosphorus Concentrations and Removal in the
Harvested Portion of Some Major Agricultural Crops

Crop category	Phosphorus concentration (%)	Crop yield (Mg/ha)	Phosphorus removal (kg P/ha)
Grains			
Corn	0.28	9.4	26
Rice	0.24	4.0	10
Soybeans	0.58	2.7	12
Wheat	0.45	2.7	12
Forages			
Alfalfa	0.25	13.4	34
Bermuda grass	0.32	22.4	71
Corn silage	0.06	67.2	39
Red clover	0.25	9.0	22
Tall fescue	0.44	9.0	39
Specialty			
Beans, snap	0.06	26.9	15
Celery	0.14	67.2	90
Onions	0.05	40.3	19
Potatoes	0.04	44.8	17
Sugarcane	0.02	224	48
Sweet corn	0.07	6.7	5
Tomatoes	0.04	67.2	27

Source: Pierzynski and Logan, 1993.

divalent (HPO_4^{2-}) anion (as illustrated in Figure 5-5b), both of which are readily available for plant uptake. One common characteristic of P in soils at all stages of weathering, however, is the low solubility of P-bearing minerals. In most agricultural soils the soil solution concentration of P ranges from <0.01 to 1 mg/L (ppm), although concentrations as high as 6–8 mg/L have been measured in recently fertilized soils. Soluble P concentrations required for normal plant growth vary somewhat between species and with yield potential. A value of 0.2 mg P/L is commonly reported as the desired concentration for soil solution P. Solution P concentrations as low as 0.03 mg/L, however, have been adequate to produce high yields of many agronomic crops. Interestingly, a soluble P concentration of 0.03 mg P/L is also used as the threshold for eutrophication in freshwaters, suggesting a commonality in P nutritional requirements between terrestrial and aquatic plants.

5.2.2 Organic Soil Phosphorus

Phosphorus dissolved from minerals enters the soil solution where biological accumulation occurs. Plants vary widely in their P contents, with most agricultural crops containing from 0.1 to 0.5% P, much less than the concentrations of other major essential elements such as N and K (2.0–4.0%). In agricultural situations some of the P accumulated by plants is removed from the soil as harvested grain or forage, with the remainder returned to the soil as crop residues (Table 5-1). In urban horticulture and native ecosystems all the P accumulated is eventually recycled back into the soil as plants die or lose vegetative material in response to normal biological development. One major environmental concern in urban areas, however, is the conversion of biologically incorporated P in private or industrial landscapes into waste organic matter (leaves and lawn clippings) that is often disposed of by municipalities through landfills or large-scale composting operations. Similarly, human wastes represent a

Table 5-2 Forms of Phosphorus in Soils Amended with Organic Wastes

	Total P	Organic P	Inorganic P
	(mg/kg)		
Manure-amended soils (Sims, 1992)			
4 Delaware soils	1467	281 (19%)	1166 (81%)
7 Pennsylvania soils	2240	427 (19%)	1813 (81%)
Pullman clay loam (Sharpley et al., 1984)			
Untreated soil	353	202 (57%)	151 (43%)
Soil + fertilizer P	457	231 (51%)	226 (49%)
Soil + feedlot waste	996	323 (32%)	673 (68%)
@ 67 Mg/ha/yr			
Greenfield sandy loam (Chang et al., 1983)			
Untreated soil	579	60 (10%)	519 (90%)
Soil + sewage sludge	1433	122 (9%)	1211 (91%)
@ 45 Mg/ha/yr			

transformation of biologically accumulated P from food chain crops to waste organic matter (sewage). An agricultural analogy to this situation is concentrated animal-based agriculture (feedlots, poultry operations) where harvested organic P is transformed into manure organic P. In both urban and agricultural systems this waste organic matter is often continually applied to soils within a limited geographic area, resulting in the buildup of soil P to levels of environmental concern, because with time much of the organic P is degraded by soil microorganisms and converted back to soluble and inorganic forms of P (Table 5-2).

Common forms of organic P found in soils include inositol phosphates, phospholipids, phosphoglycerides, phosphate sugars, and nucleic acids. Microbial decomposition of organic P results in the release of soluble P, which with time is normally converted into stable inorganic forms of P. Studies have shown, however, that as much as 50% of the P transported in runoff can be soluble organic P. The extent and rate of conversion of organic P into soluble or stable inorganic forms are highly dependent on the nature of the original organic material, as well as environmental factors such as pH, temperature, and soil moisture. Fresh plant residues may quickly release P into the soil solution where more stable forms of organic matter (such as soil humus, animal manures, sewage sludges, or composts) generally act as long-term, slow-release sources of P. Mineralization — the conversion of organic P to inorganic P — usually occurs rapidly if the C:P ratio of the organic matter is <200:1, while immobilization — the incorporation of P into microbial biomass — occurs if C:P ratios are >300:1. At this time, it is unclear whether plants can directly absorb soluble organic P or further enzymatic action is required to cleave phosphate molecules from the organic compound prior to uptake.

5.2.3 Phosphorus Additions to Soils

The low solubility of P in many soils means that the soil solution must be replenished frequently and rapidly with soluble P if adequate plant growth is to occur. In soils with high fertility levels, the soil alone can normally provide all the P required by plants. The need for additional P can be easily determined by *soil testing*. In essence, soil testing involves comparison of available soil P, as estimated by a chemical extractant, with *critical values* determined from long-term research (Figure 5-6). As the soil test P value of a field becomes progressively lower, the likelihood

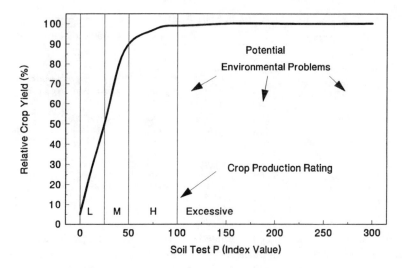

Figure 5-6 The relationships between soil test P, crop yield, and potential for environmental problems due to excessive soil P (L = low, M = medium, H = high).

of an economically profitable crop yield response to P fertilization increases. Similarly, in fields that have soil test values above the critical level, addition of P fertilizers will in all likelihood not result in crop yield increases. Application of fertilizers, manures, or other P sources to fields with soil test values well above the critical level is unnecessary and increases the possibility of soluble or sediment-bound P losses to surface waters. One exception to this has been the use of small amounts of "starter" fertilizer placed near the seed or seedling at planting, a practice that has frequently been shown to produce more vigorous early plant growth and to occasionally increase crop yields (Figure 5-7). Plant response to starter fertilizers, even in high P soils, is often explained by the lower rates of mineralization of organic P and diffusion of soluble P that occur early in the growing season when cool soil temperatures are common. As soils warm, microbial and chemical reactions occur more rapidly and the supply of P to plant roots increases to acceptable levels.

Many soils, however, due to low P parent material, low organic matter contents, or intensive cropping do not contain adequate P for the plant growth desired. Fertilization with inorganic P fertilizers or the use of manures, sludges, composts, or other waste products is then necessary. Commercial P fertilizers are produced by industrial processes that involve the reactions of phosphate ores (e.g., fluorapatite, or "rock phosphate") with sulfuric acid (ordinary superphosphate, 9% P), phosphoric acid (triple superphosphate, 20–22% P), or phosphoric acid with ammonia (monoammonium phosphate, 21% P; diammonium phosphate, 23% P). Rock phosphate (13–17% P) can also be used directly as a P fertilizer although its low solubility often means high application rates are needed to provide sufficient available P. The P content of the most commonly used inorganic and organic sources of P is provided in Table 5-3.

Animal manures, municipal sewage sludges or composts commonly applied to agricultural lands as part of waste disposal operations contain relatively low

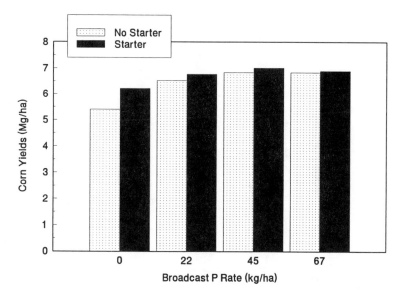

Figure 5-7 The effect of starter fertilizer on corn yield. Results of a 20-year study in Iowa comparing the use of starter P applied with differing rates of broadcast P at 3-year intervals. (Adapted from Young et al., 1985.)

concentrations of P (0.1–3.0%). However, the large and frequent applications normally used with these organic wastes, in combination with the relatively small amounts of P actually removed from the soil in harvested grain or forage, not only can provide for crop P requirements, but also (as mentioned above) can build soil P to extremely high levels within a few years (Figure 5-8). Unfortunately, because of transportation and handling costs, most animal-based agricultural operations have few alternatives to land application of manures within a short distance from the area where the animals are grown. This often results in the long-term buildup of soil P to excessive levels as typical manure application rates usually add P at rates beyond crop P requirements. If these soils are near a sensitive water body, the potential environmental impacts of P become an issue. Similarly, for municipalities the alternatives available to land application of organic wastes (landfills, ocean dumping, and incineration) are increasingly limited and costly, and may have more serious environmental impacts than well-managed land application programs.

Inorganic wastes from industrial operations such as basic slag (3–5% P) — a steel manufacturing by-product — or coal ash (<1% P) — a by-product of the electric power industry — can add considerable amounts of total P to soils. While the solubility and thus plant availability of industrial wastes are normally less than commercial fertilizers and animal manures, land application is increasingly becoming an economically desirable alternative to landfill disposal of these materials. Management programs that use these materials as a soil amendment must also consider their effects on soil pH and soluble salts, the presence of potentially phytotoxic elements (e.g., B, Na), and the long-term fate of nonessential heavy metals found in these by-products.

Table 5-3 Major Inorganic and Organic Sources of Phosphorus for Crop Production

Phosphorus source and chemical composition		%P	%P$_2$O$_5$	Other nutrients
Commercial fertilizers				
Ordinary superphosphate	[Ca(H$_2$PO$_4$)$_2$ + CaSO$_4$)]	7–10	16–23	Ca, S (8–10%)
Triple superphosphate	[Ca(H$_2$PO$_4$)$_2$]	19–23	44–52	Ca
Monoammonium phosphate (MAP)	[NH$_4$H$_2$PO$_4$]	26	61	N (12%)
Diammonium phosphate (DAP)	[(NH$_4$)$_2$HPO$_4$]	23	53	N (21%)
Urea-ammonium phosphate	[CO(NH$_2$)$_2$,NH$_4$H$_2$PO$_4$]	12	28	N (28%)
Ammonium polyphosphates (liquid)	[(NH$_4$)$_3$HP$_2$O$_7$]	15	34	N (11%)
Rock phosphates				
U.S. (Florida)	[Ca$_{10}$F$_2$(PO$_4$)$_6$ · XCaCO$_3$]	14	33	Major impurities:
Brazil	(varies between mineral deposits)	15	35	Al, Fe, Si, Fe, CO$_3^{2-}$
Morocco		14	33	
Former U.S.S.R.		17	39	
Organic phosphorus sources				
Beef manure		0.9	2.1	N, K, S, Ca, Mg,
Dairy manure		0.6	1.4	and microelements
Poultry manure		1.8	4.1	
Swine manure		1.5	3.5	
Aerobically digested sludge		3.3	7.6	
Anaerobically digested sludge		3.6	8.3	
Composted sludge		1.3	3.0	

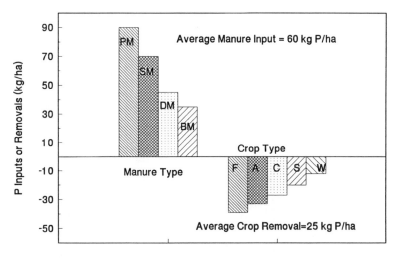

Figure 5-8   Comparison of typical inputs of P from animal manures and crop removal in harvested grain or forage. (PM = Poultry manure, SM = swine manure, DM = dairy manure, BM = beef manure; F = fescue, A = alfalfa, C = corn, S = soybeans, and W = wheat.)

5.3 PHOSPHORUS TRANSFORMATIONS IN SOILS

The ultimate goal of soil P management is to maintain the concentration of soil solution P at a value adequate for plant growth, while minimizing the movement of soluble P, organic P, and sediment-bound P to sensitive water bodies. Proper soil, crop, and water management practices — such as fertilization techniques and timing, crop rotation, tillage, irrigation, and drainage — that maximize the efficiency of P recovery from fertilizers and wastes represent the practical applications of the basic knowledge gained from an understanding of the soil P cycle.

For crop production purposes these practices focus on ensuring a rapid replenishment of the soil solution to optimize P nutrition and on minimizing soil loss to prevent long-term degradation of the soil resource and thus soil productivity. Although similar in many respects to agricultural practices, environmental management practices for soil P must be more intensive and have less room for error due to the low levels of P required to induce eutrophication. In either case, basic knowledge of the chemical, microbiological, and physical transformations undergone by soil P is necessary to develop innovative approaches that accomplish both goals. The major soil inorganic P transformations of importance include the fixation of P in insoluble forms by *adsorption* and *precipitation* reactions, and the solubilization of P by *desorption* reactions and mineral *dissolution*. Soil organic P transformations are primarily *mineralization-immobilization* reactions mediated by soil microorganisms and P uptake by plant roots alone or in association with mycorrhizal fungi.

5.3.1 Adsorption and Desorption of Soil Phosphorus

Adsorption refers to the removal of ionic P ($H_2PO_4^-$, HPO_4^{2-}) from solution by a chemical reaction with the solid phase of the soil. It is a general term that refers to

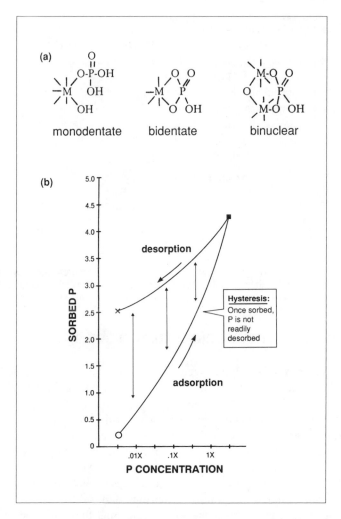

Figure 5-9 Phosphorus adsorption and desorption processes in soils. (a) Proposed mechanisms of adsorption of P by soil constituents (M represents a metal [Fe, Al] ion in
a clay or oxide). (b) Idealized adsorption-desorption curve for P, illustrating hysteresis. When P is added to the soil, as in fertilization, and the concentration of P in
the soil solution (x) increases (e.g., from 0.01x to 1x), adsorbed P increases as
well. However, there is a the lack of complete reversibility in desorption of P from
the solid phase, referred to as hysteresis, when the concentration of P in the soil
solution is decreased from 1x to 0.01x by plant uptake or leaching. (Adapted from
Fixen and Grove, 1991. With permission.)

the formation of a chemical bond between the phosphate anion and a soil colloid, but
does not specify the mechanism of P retention, although several types of bonding
have been proposed (Figure 5-9a). The primary reactive phases in soils and sediments
responsible for adsorption reactions are clays, oxides or hydroxides of Fe and Al,
calcium carbonates, and organic matter. Adsorbed P is normally considered to be
slowly available and capable of gradually replenishing the soil solution in response
to plant uptake. The term "labile" soil P is often use to describe that portion of

adsorbed P that will be readily available for plant uptake and can be easily extracted and measured, along with soluble P, by a chemical soil test.

Desorption refers to the release of P from the solid phase into the solution phase. Desorption occurs in soils when plant uptake depletes soluble P concentrations to very low levels, or in aquatic systems when sediment-bound P interacts with natural waters with low P concentrations. Only a small fraction of adsorbed P in most soils is readily desorbable. Much of the P that is added in fertilizers or manures is rapidly and irreversibly "fixed"; and is not subject to desorption, a phenomenon referred to as *adsorption hysteresis* (Figure 5-9b). Hysteresis reflects the transformation of soluble P into very unavailable forms by processes such as precipitation of P as insoluble compounds, occlusion of P by other precipitates (e.g., Al and Fe oxides), and slow diffusion of P into solid phases in the soil. Studies have shown that, while the rate of P desorption is initially quite rapid, it often decreases markedly within a short time. From an environmental perspective this suggests that short, intense rainfall events may quickly deplete the desorbable fraction of soil P and redistribute this P into runoff.

Soils and sediments differ widely in their capacity to adsorb and desorb P. In general, highly weathered soils will adsorb more P due to their greater clay and Al and Fe oxide contents. As shown in Figure 5-10, the amount of P adsorption that must occur to attain a P concentration in solution adequate for plant growth (0.2 ppm) would be ~2500 mg P/kg for a Hawaiian inceptisol with a 70% clay content, compared to ~50 mg P/kg for a Peruvian ultisol with a 6% clay content. Adsorption also varies markedly with soil depth and is affected by cultural operations that alter available soil P levels; soil pH; and organic matter content, such as fertilization, liming, manuring, or reduced tillage operations. The importance of subsoils in P adsorption and leaching is shown in Figure 5-11 for two sandy, Atlantic coastal plain soils. The Evesboro soil, which has sandy surface and subsoil horizons, has much less adsorption capacity throughout the soil profile than the Matawan sandy loam, which has a subsoil horizon with 60% clay. The higher soil test P (STP) values in the Evesboro subsoil reflect the greater mobility of P in profiles with little accumulation of clay. Tillage practices that affect the distribution of P can also alter the adsorption and desorption of P. No-tillage agriculture, where P is not incorporated by plowing, often results in considerable stratification of P in the soil, with extremely high P values in the top few centimeters. As shown in Figure 5-12, this can result not only in a much lower P adsorption capacity in this depth (0–2 cm), but also in the presence of very high levels of easily extractable P relative to the remainder of the topsoil horizon (6–18 cm), a situation that could contribute to high levels of soluble P in runoff.

In general, soils that are low in P, acidic, and high in clay or Fe and Al oxides — particularly noncrystalline oxides — have the greatest P adsorption capacities. In calcareous soils, $CaCO_3$ and Fe oxides are the dominant factors controlling P adsorption. Sandy soils have low P adsorption capacities and are the most susceptible to leaching of P, as are soils with extremely high organic matter contents (peats, heavily manured soils, etc.) where soluble organic matter may enhance the mobility of P by coating surfaces responsible for P adsorption. Additionally, organic P complexes have been shown to leach more rapidly and to greater depths than

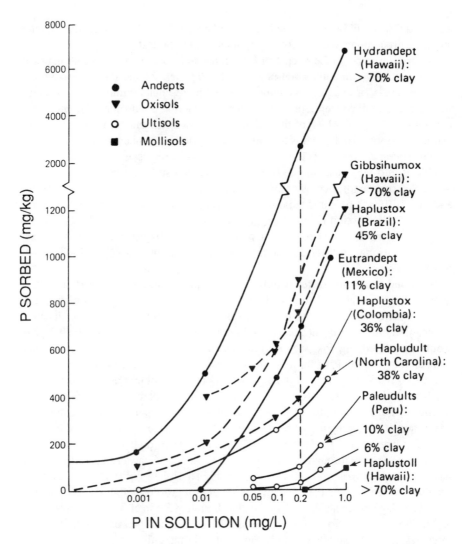

Figure 5-10 Differences in P adsorption capacity of soils from four soil groups. Vertical dashed line represents the amount of P that must be adsorbed to attain a level in solution adequate for the growth of most crops (0.2 mg/L). (Adapted from Sanchez and Uehara, 1980. With permission.)

inorganic soluble P, raising questions about the subsurface movement of P into surface waters by lateral groundwater flow, particularly in artificially drained soils.

Adsorption of P by soils can be quantified by the use of mathematical equations that relate the amount of P retained by a solid phase (P_{Ads}) to the concentration of P in solution at equilibrium, when adsorption reactions have ceased (P_{eq}). Data are usually obtained by reacting soils or sediments with solutions ranging in initial P concentration from 0 to 20 mg P/L. Plots of the quantity of P adsorbed vs the equilibrium P concentration are referred to as "adsorption isotherms," as illustrated

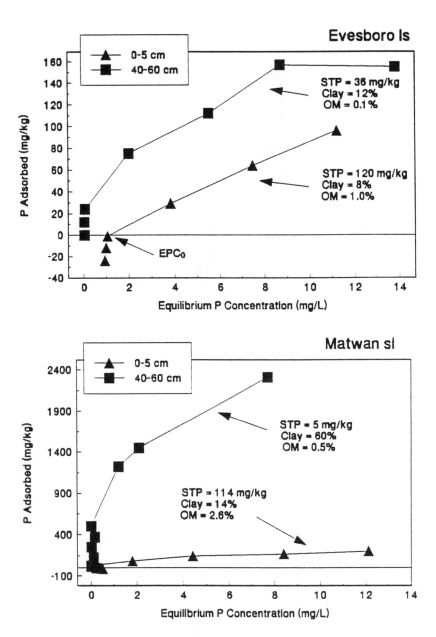

Figure 5-11 Phosphorus adsorption isotherms for two depths of two Atlantic coastal plain soils. The 0–5-cm depth represents the zone of maximum interaction between runoff and soil P, and the 40–60-cm depth represents the potential for subsoils to retain P against leaching. STP refers to value of soil test phosphorus prior to adsorption (soil test = Mehlich, 1, 0.05 N HCl + 0.025 N H_2SO_4). EPC_0 is equilibrium concentration of P in solution when the rate of P adsorption and desorption are the same. (Sims, unpublished data.)

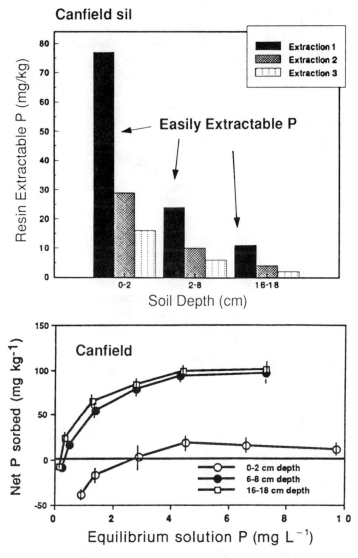

Figure 5-12 Comparison of easily extractable P and P adsorption isotherms for upper and lower portions of the topsoil of a Canfield silt loam. (Adapted from Guertal et al., 1991. With permission.)

in Figures 5-11 and 5-12. The most commonly used approaches are the Langmuir and Freundlich equations, expressed mathematically as follows:

$$P_{Ads} = \frac{kP_{eq}b}{[1 + kP_{eq}]} \qquad (5\text{-}1)$$

$$P_{Ads} = kP_{eq}^{1/n} \qquad (5\text{-}2)$$

In the Langmuir equation the parameters k and b are presumed to characterize the energy of bonding of P with the solid phase (k) and an estimate of the maximum amount of P that can be adsorbed (b). For the Freundlich equation, k and n are empirical constants that vary according to soil properties. Although these equations do not provide mechanistic information about the processes of P adsorption, they are useful in comparing adsorption capacities between soils. They are also used to approximate the quantity of P that must be adsorbed by various soils to raise the P concentration in the soil solution at equilibrium to a desired, or maximum, value.

Another useful parameter than can be obtained from P adsorption isotherms is the "EPC_0" or equilibrium phosphorus concentration at zero sorption. The EPC_0 value represents the P concentration maintained in solution by a solid phase (soil, sediment) when the rates of P adsorption and desorption are the same. EPC_0 values are used to characterize the potential of surface soils, stream sediments, and stream bank materials in contact with runoff or stream flow to remove P from or release P into the flowing waters. Soil particles or sediments with high EPC_0 values that contact low P waters (rainfall, subsurface discharge) will desorb P into the waters; conversely, if the solid phases have low EPC_0 values, they can reduce the P concentration of stream flow or runoff and thus decrease the potential for downstream eutrophication. The high EPC_0 values for the top few centimeters of the Evesboro soil shown in Figure 5-11 (~1 mg/L) and the Canfield soil shown in Figure 5-12 (~3 mg/L) indicate that both of these high P soils should readily desorb soluble P into runoff waters.

5.3.2 Precipitation of Soil Phosphorus

Precipitation can be defined as the formation of discrete, insoluble compounds in soils, and can be viewed as the reverse of mineral dissolution. The most common forms of precipitated P include the products of reactions between soluble, ionic P and Ca, Al, and Fe. In calcareous soils, where the pH is high and soluble Ca is the dominant cation, the addition of soluble P initially results in the formation of dicalcium phosphate dihydrate ($CaHPO_4 \cdot 2H_2O$), which with time slowly reverts to other more stable Ca phosphates such as octacalcium phosphate [$Ca_8H_2(PO_4)_6 \cdot 5H_2O$], and in the long term to apatite [$Ca_{10}(PO_4)_6F_2$]. The chemistry of calcium phosphates is reasonably well understood and has been studied by the use of mineral stability diagrams that, based on chemical equilibria, can predict the mineral that controls soil P solubility. Examining the effect of cultural practices or soil properties on changes in these stability diagrams can help one to understand the long-term cycling of P in soils. In acidic soils, where soluble Al and Fe are the major soluble cations, Fe and Al phosphates are the dominant precipitates. The presence of amorphous or partially crystalline Fe and Al oxides in these soils that can "occlude" P as they crystallize, makes identification of discrete solid phases of Fe-P and Al-P difficult. Dissolution of precipitates in calcareous and acidic soils is highly pH dependent. The solubility and thus availability of P to plants (and runoff or leaching waters) are generally greatest under slightly acidic conditions (pH 6.0–6.5). Application of the principle of chemical precipitation to remediation of eutrophic lakes has been successful and commonly involves addition of alum (aluminum sulfate) to the epilimnion (upper portion of the water column) to precipitate P as aluminum phosphate. In addition to direct precipitation of soluble P, the alum forms a layer of aluminum hydroxide on the lake bottom that is highly reactive for P subsequently desorbed from lake sediments.

5.3.3. Mineralization and Immobilization of Soil Phosphorus

Organic P in soils, or P added to soils in organic wastes or crop residues, represents an important source of P for plant growth, as well as a potential source of soluble P that may be lost in runoff or leaching. Mineralization refers to the microbially mediated decomposition of organic compounds, resulting in the release of inorganic forms of nutrients into the soil solution. Immobilization, conversely, is defined as the "tieup" of mineral elements such as P by soil microorganisms into biochemical compounds essential for microbial metabolism. Mineralization of organic P is largely controlled by the relative amount of carbon in the organic substrate, which acts as the energy source for the decomposing microorganisms. High C:P ratios provide substantial energy and stimulate microbial growth that consumes all available P; low C:P ratios can result in excess soluble P beyond microbial needs that is then available for plant uptake. Environmental conditions, as would be expected, play a key role in the rate and extent of decomposition of organic P compounds. Optimum conditions for mineralization of organic P are similar to those for soil organic N (see Chapter 4). Once mineralized from organic matter, P reverts rather quickly to inorganic forms such as adsorbed or precipitated P. In many instances frequent, long-term additions of animal manures and other organic wastes to soils have been shown to primarily increase total inorganic P in soils, not total organic P (Table 5-2).

5.4 PHOSPHORUS TRANSPORT IN THE ENVIRONMENT

All forms of P — soluble, adsorbed, precipitated, and organic — have been shown to be susceptible to transport from source soils to water bodies. The fact that the greatest reservoir of P in the earth's environment is ocean sediments is proof that "leakage" of P from geologic sources is a continuous process. Controlling this transport is necessary to avoid the undesirable effects associated with eutrophication and to avoid the waste of natural resources that occurs when geologic deposits of P or reclaimed waste organic P (e.g., manures, sludges) are used as soil amendments to improve plant growth and yield. Transport of soil P occurs primarily via surface flow, although the background levels of P entering streams and lakes via subsurface flow certainly reflect the impacts of land use. Direct discharge of P from wastewater treatment systems into rivers represents a distinctly different form of transport that requires enormous technological and economic costs to reduce P effects on the aquatic environment.

5.4.1. Transport of Phosphorus by Surface Flow

Water flowing across the soil surface can dissolve and transport soluble P, or erode and transport particulate P. Soluble P can be either inorganic or organic, while particulate P generally consists of finer sized soils particles (e.g., clays) and lighter organic matter. The small quantity of soluble or readily desorbable P in most soil environments is due to the low solubility of P, the considerable adsorption capacities of clays for P, and the high concentration of P in soil organic matter. This results in the majority of total P transport occurring as particulate P. This is particularly true where runoff contains high quantities of suspended solids. However, if natural or artificial filtration processes are operational (e.g., forests, wetlands, grassed waterways) to remove sediment-bound P, then the transport of soluble P becomes of more importance. Similarly, if soils are excessive in P, such as often occurs in areas where organic wastes are frequently applied in agricultural operations, the level of soluble P is of greater importance.

Virtually all soluble P transported via runoff is biologically available, but particulate P that enters streams and other surface waters must first undergo some type of solubilization reaction (e.g., desorption) before becoming available for aquatic biota. Complicating the matter further is the fact that during the transport process itself, soluble and particulate P can undergo reactions with soils bordering fields and lakes, flowing waters, and stream bed and stream bank sediments. These reactions can dramatically alter the potential for the P originally lost in runoff to induce eutrophication downstream. Soil management practices seek to reduce P losses in runoff by decreasing excessive soluble soil P and reducing erosive loss of fine-sized particles and organic matter. Nutrient management programs that address fertilizers and organic wastes, conservation tillage practices, and erosion control mechanisms such as terraces and grassed waterways are examples of strategies used to minimize P loss in runoff.

The mechanisms involved in soluble P transport are straightforward, and include an initial desorption or dissolution of P bound by soil particles, followed by water movement from the source soil to a stream or river that later intercepts a sensitive water body. Soluble organic P that is not adsorbed by soil particles may also be carried by runoff. Of importance from a soil management point of view is the fact that most of the soluble P entering runoff originates from interaction of flowing water with the uppermost surface (0–2 cm) of the soil. Any management practice that minimizes soluble P levels at this depth (e.g., fertilizer or manure incorporation) or enhances infiltration of water into the soil, where soluble P can be adsorbed by soil constituents, will decrease P loss to the environment. In a similar manner, the natural filtration capacity of nonagricultural areas bordering fields or the high sorption capacities of stream and stream bank sediments can reduce P transport by adsorbing P from runoff waters, assuming adequate time of contact between P-enriched waters and sediments.

Some examples of methods used to reduce P losses in runoff are provided in Table 5-4 which shows the value of reduced tillage practices (chisel plow or no-till vs conventional), crop residue strips, and grassed waterways in trapping sediment-bound and soluble P. These examples also illustrate the importance of considering all aspects of runoff control practices to maximize the overall efficiency of a management system. Chisel plowing, for example, may be just as effective in controlling erosion and runoff as complete no-tillage in certain soils. Chisel plowing also improves crop yield and nutrient uptake by breaking apart compaction zones and improving root growth; it also provides a method to at least partially incorporate organic wastes. Narrower crop residue strips may be just as effective as wider strips in reducing P loss, if adequate soil cover is present. Multiple cutting practices for

Table 5-4   Examples of the Methods Used to Control Phosphorus Loss in Runoff

Tillage Practices for Corn (Andraski et al., 1985)						
	Year:	1980	1981		1982	1983
Tillage method	Month:	September	June	July	October	June	July
		Total P loss in runoff (mg/m³)					
Conventional		133	8	230	220	175	22
Chisel plow		21	<1	20	37	67	11
No-till		39	<1	20	24	21	5

Cornstalk Residue Management (Alberts et al., 1981)				
Strip width (m)	% Cover	Available P (g/hr/m of width) Entering	Leaving	Reduction in P Loss (%)
1.8	27	0.98	0.77	21
1.8	50	1.39	0.70	50
2.7	50	2.63	0.48	82
4.6	50	1.80	0.51	72

Wastewater Renovation by a Reed Canary Grass Filter Strip (Payer and Weil, 1987)		
	P Removal efficiency from wastewater (%)	
Cover crop management	Dissolved P	Total P
Multiple cuttings, harvest crop residue	32	70
Multiple cuttings, leave crop residue	20	62
Cut once, leave crop residue	4	50

grassed waterways increased the efficiency of P removal in an overland flow waste-water renovation system. However, equipment damage to the grassed waterway caused by the extra cuttings produces ruts and channels that could contribute to increased runoff. In each of these examples, the need to reduce P loss must be balanced against other aspects of the system (e.g., crop growth in severely compacted soils, loss of cropland by conversion to filter strips, sustainability of grassed water-way) to optimize the efficiency of the management practice.

The transport and subsequent reactions of particulate P are considerably more complex. Particulate P originates not only from soil erosion, but also from the beds and banks of streams or drainage areas that carry water from a field to a surface water body. As with all erosion, the loss of particulate P represents the detachment of soil particles as a result of the energy contained in falling or flowing water. The nature of the rainfall or runoff event will determine the energy level of the water, while the texture and structure of the soil will affect the amount of energy required to detach and transport soil particles. As smaller, lighter particles require less energy to dislodge and carry, it is not surprising that clays and organic matter — soil constitu-ents with relatively low densities — are preferentially transported in runoff. Once soil solids enter a watercourse, they eventually settle from the water as sediment. Resuspension of these sediments during intense rainfall events or rapid stream flow, along with detachment of soil particles from stream banks, can increase particulate P transport at a later date. In general, more intensive land use practices — such as urban construction activities, surface mining, and conventionally tilled agricultural soils — will have a greater percentage of particulate P transport from surface runoff, relative to resuspended sediments and stream bank materials. In natural ecosystems and areas where soil conservation practices decrease erosion of soil solids, the role of streams in the transport of particulate P is of more importance.

Table 5-5   Average Enrichment Ratios for Six Western Soils

Soil property	Enrichment ratio[a]
Organic carbon	2.00
Clay (%)	1.56
Surface area	1.32
Bioavailable P	1.49
Total P	1.46
EPC (mg/L)	−1.80[b]

Source: Adapted from Sharpley, 1985.

[a] $\text{Enrichment ratio} = \dfrac{[\text{Value of soil property in runoff sediment}]}{[\text{Value for soil property in source soil before runoff}]}$

[b] Negative enrichment ratio indicates that source soil maintained a higher P value in solution at equilibrium than runoff sediment.

The particles transported in runoff are normally higher in P, organic matter, and other nutrients, relative to the soil from which they originated. This enrichment occurs due to the greater sorption capacity of the finer sized clay particles and the relatively high concentrations of P in organic matter, relative to silt and sand particles (Table 5-5). The "enrichment ratio" concept was developed to identify watersheds where more intensive management practices are required to control nutrient enrichment of surface waters. The "enrichment ratio" is calculated as the ratio of the concentration of P in the particulate matter found in runoff to that in the source soil. Research has shown that enrichment ratios for total and available P are closely related to the amount of soil lost from a watershed. This provides a quantitative basis for estimating the impact of land use on the loss of soluble, bioavailable, and particulate P and thus productivity of soils and surface waters in a watershed. Recent advances in research have resulted in models that have been remarkably successful in predicting P loss from readily obtainable parameters such as soil loss from erosion, soil test P, and soil properties associated with the range of values for soil series found in the landscape (e.g., clay content, organic matter, soil total P) (Figure 5-13).

Once transported, particulate P has entered a lake, many of the processes operating in the soil P cycle continue, albeit at different rates. The size of the particles and the relative temperatures of the inflowing water and the lake influence the initial distribution of the sediments within the lake. One of the major causes of eutrophication is the high degree of turbidity associated with fine-sized particles that settle very slowly to the lake bottom. The turbid nature of the upper water column reduces light penetration, decreasing the photic zone. Because the finer sized particles are usually higher in P than the quickly settling coarser particles, the overall initial effect is to stimulate biological productivity in the epilimnion. Of long-term importance, however, is the later desorption of P from sediments that with time have settled to the bottom of the lake. The P adsorbed by clays, oxides of Fe and Al, and $CaCO_3$ are the main sources of potentially desorbable P. The physical and chemical processes operational in the lake greatly influence the desorption process. Changes in temperature or turbulence induced by storms can resuspend bottom sediments in the water column and enhance the desorption of P in "photic" zones where algal growth occurs. The oxygen status of the lake can also affect desorption of P from sediments. Studies have shown, for instance, that under anaerobic conditions P desorption is enhanced,

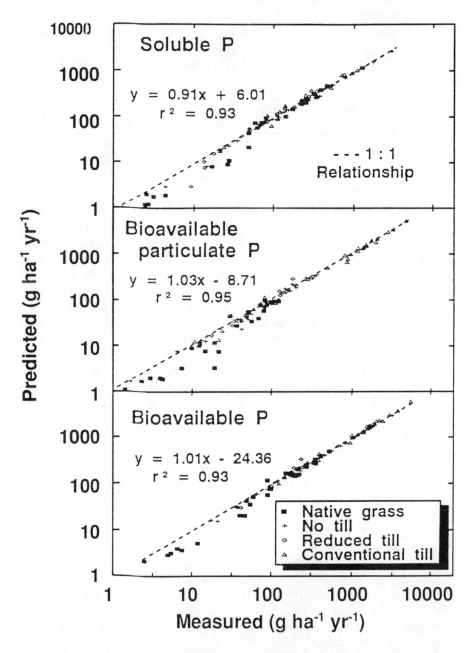

Figure 5-13 Comparison of predicted and measured losses of P. Soluble P was predicted from soil loss, soil test P, and variables related to runoff. Bioavailable P was predicted from soil total or bioavailable P, sediment concentration in runoff, and enrichment ratios estimated from soil loss. (Adapted from Sharpley and Halvorson, in press.)

and that when lake oxygen levels increased, P sorption increased as well. The residence time of water in a lake also greatly affects the importance of P in sediments. When inflowing water has a long residence time (at least several months), the P inputs from streams and rivers have sufficient time to accumulate as sediments on lake bottoms. If water flow through the lake is rapid, however, P retained by finer sized particles may be removed from the lake prior to settling from the water column. Indeed, one common strategy used in urban and agricultural situations to minimize nutrient enrichment of lakes and reservoirs, is to construct settling ponds or impoundments with long residence times that act as filtration systems for particulate P in runoff. Subsequent removal or chemical "sealing" of the enriched sediments will be needed, but can be anticipated and planned for maximum efficiency.

5.4.2. Phosphorus Leaching and Subsurface Flow

The natural process of groundwater discharge into surface waters contributes P that can play a role in eutrophication. In most instances the input of P by subsurface flow represents the effect of decades of land use and must be viewed as a base flow contribution that can be altered only by long-term improvements in management of all urban and agricultural operations in a watershed. One possible exception is the subsurface discharge from artificially drained fields, where corrective measures can be taken either at the source or at the point of discharge. Improved nutrient management programs and artificial wetlands are examples of strategies that can be used to reduce nutrient loads in drainage waters from urban and agricultural areas.

In most cases the concentrations of P in subsurface flow have been found to be quite low and well below the eutrophication threshold. This reflects the considerable sorption capacity of soils for P, particularly P-deficient subsoil horizons. Because of this, P leaching, unlike nitrate leaching, is rarely viewed as an important environmental issue. Exceptions include organic soils with fluctuating water tables; and heavily manured, sandy soils with shallow water tables. The role of organic matter in P leaching (mentioned above) is not well understood, but is believed to be a key factor. Artificial drainage — commonly used in organic soils — normally increases infiltration and percolation of water, increasing the likelihood of P leaching, but decreasing P losses in runoff. If heavy applications of manure are combined with artificial drainage of soils, P leaching can be significant (as shown in Figure 5-14) where the application of 200 Mg/ha of dairy manure sharply increased dissolved P concentrations in tile drainage. Improved aeration in organic soils due to artificial drainage has also been shown to stimulate P mineralization and loss in drainage waters via leaching.

The most common situations where P leaching from mineral soils have been observed are in municipal and agricultural wastewater treatment systems and sandy soils where manure disposal operations have been necessitated by shortages of suitable land for proper manure management. Examples of this are shown in Figure 5-15 for sandy soils used for a wastewater irrigation system and for agriculture crops in an area dominated by a highly concentrated poultry industry. In both cases, P movement into subsoils has clearly occurred. In the case of the wastewater irrigation system, groundwater P values in the sprayed areas averaged 0.75 mg/L, relative to a local environmental standard of 0.05 mg/L.

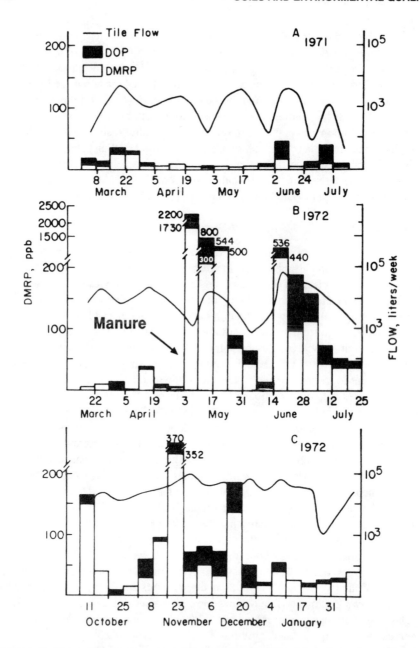

Figure 5-14 Effect of dairy manure application (200 Mg/ha) and tile flow on the loss of dissolved organic P (DOP) or dissolved soluble P (DMRP) in tile drainage during a 2-year period. (Adapted from Hergert et al., 1981. With permission.)

5.5 ENVIRONMENTAL MANAGEMENT OF SOIL PHOSPHORUS

Management programs to minimize environmental damage from soil P must be multidisciplinary in scope. Because eutrophication is the only negative environmental

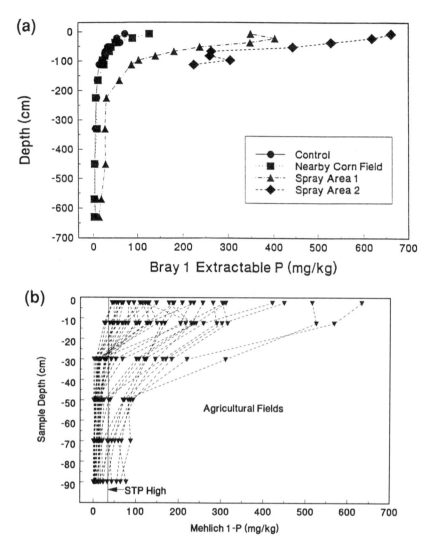

Figure 5-15 Phosphorus leaching studies illustrating distribution of soil test P with depth (soil tests: Bray P [0.03 N NH$_4$F + 0.025 N HCl]; M1-P [Mehlich 1, 0.05 N HCl + 0.025 N H$_2$SO$_4$]). (a) A wastewater irrigation system compared to a commercially fertilized corn field. (Adapted from Adriano et al., 1975.) (b) From Delaware 34 agricultural soils where animal manures are regularly applied. (Mozaffari and Sims, 1993.).

effect of P, the strategies to control P pollution are conceptually simple. Soil P levels should be maintained at levels adequate, but not excessive for plant growth; and runoff of soluble and particulate P must be minimized by proper soil and water management. In water bodies where eutrophication has already occurred, remediation involves practices that decrease the solubility of P and desorption of P in the water body. To date, no systematic approach that controls P loss during the transport process has been developed, although the use of controlled drainage systems and artificial wetlands are recent advances in this area.

Many significant obstacles exist to implementation of these "simple" practices, however. Foremost is the availability of human and economic resources needed to design and implement improved management programs for reducing soil loss in urban and agricultural areas. Another significant limitation, particularly in urban areas or areas where animal-based agriculture is predominant, is the continued concentration of P from other geographic areas into organic (or inorganic) wastes. The increasing costs of waste transport, treatment, and landfilling frequently result in the application of these wastes to soils in a manner that creates environmentally excessive levels of soil P. Given the formidable nature of these limitations to soil P management, it is apparent that a systematic, long-term approach will be required. Multidisciplinary teams with expertise in aquatic ecology and chemistry, hydrology, soil science, crop production, urban and industrial waste management, and resource economics must identify the areas requiring immediate attention and the broadly applied, long-term solutions. Strategies required for urban and agricultural situations while different, must be coordinated, given the interdependence of farm and municipal populations. Computer-based water quality models have been developed that focus on P management, such as erosion-productivity-impact-calculator (EPIC). These models require a large data base with many parameters, but can be very successful in predicting changes in soil P levels or crop yield responses (Figure 5-16).

Simpler approaches that can be used by advisory agencies (Cooperative Extension, Soil Conservation Service) or crop consultants are also under development. One example of a field-oriented approach to predicting the potential environmental impact of soil P is the P index system (Figure 5-17) which uses eight readily obtainable site characteristics to obtain an index of site vulnerability to P loss. Simple or complex, the critical features of these programs are minimizing accumulations of P to excessive levels, and managing soil and water to reduce runoff and erosion.

5.5.1 Management of Soil Phosphorus: Soil Testing and Nutrient Budgets

Agricultural operations include agronomic crop production, animal production, vegetables and specialty crops, and commercial horticultural operations (such as greenhouses, nurseries, and turf farms). Adequate P fertility is necessary for maximum economic plant production in all these operations, but in many situations poor nutrient management has resulted in excessive use of P fertilizers or manures, with resultant losses as soluble P or of enriched sediments. Phosphorus management for agricultural purposes requires the development and implementation of a nutrient management program that focuses not only on profitability, but also on minimizing environmental impacts of P. Key aspects of this program will include realistic assessments of plant P requirements, soil testing and plant analysis programs that monitor soil P levels, timely and efficient P fertilization practices, cultural operations that reduce soil loss, and sound approach to organic waste management.

Although P is an essential plant nutrient with numerous key biochemical functions, the quantity required by plants is much less than the other macronutrients (N, K). Typical concentrations of P in crops range from 0.1 to 0.5%, and P fertilizer recommendations usually range from 10 to 100 kg P/ha. Crop uptake, however, is distinct from removal of P from the field in harvested plant material, as was shown in Figure 5-8. The nature of the crop rotation and overall farming operation will ultimately determine the long-term soil nutrient "budget." For instance, much of the fertilizer P taken up by agronomic grain crops is returned to the soil in the form of crop residues, while forage or silage crops remove more P from the field. Similarly, the contrasting scenarios illustrated in Figure 5-18 show the differences in complexity of P management for a cash grain farm compared to a livestock farm.

Nutrient budgets must be viewed from a state or regional perspective as well, especially in areas dominated by animal-based agriculture or in heavily urbanized regions where land application of municipal organic wastes is an accepted practice. Formidable problems exist in certain areas, both at the farm and state level as illustrated in Table 5-6 for Delaware, the site of the most concentrated poultry production industry in the United States.

Soil testing and plant analysis have traditionally been used to identify P deficiency and the likelihood of crop response to P fertilization. In many areas of the United States (as shown in Figure 5-19), P deficiencies are common and soil testing plays a vital role in ensuring maximum economic crop yields. In other areas, however, the routine use of P fertilizers and organic wastes has built soil test P levels to such high values that it may be years before economic response to P fertilization occurs (with the possible exception of P in "starter" fertilizers). An example of the long "drawdown" period required for high P soils is given in Figure 5-20, where ~15 years were required to reduce soil test P from ~100 mg P/kg to a yield limiting level (~20–25 mg P/kg). New soil testing procedures are being investigated for these P-enriched areas to identify the long-term P-supplying capacity of the soils and the need for more intensive soil management strategies due to the high levels of potentially bioavailable P present in the soils. For example, until recently many soil testing laboratories, due to equipment or economic limitations, did not measure the actual concentration of P in soils beyond some "high" value where crop response was no longer likely. The advent of environmental concerns about soil P and new instrumentation have made routine measurement of actual soil P values possible, and in many cases have identified farming operations with levels far in excess of the amount of P required for normal plant growth. As an example, a survey conducted on 70 farms in southeastern Pennsylvania, an area dominated by animal-based agriculture, found Bray P1 soil test values ranged from 36 to 411 mg P/kg and averaged 130 mg P/kg. The critical level for Bray P1 was 30 mg P/kg. Other monitoring approaches under evaluation include use of routine soil tests to assess the amount of "bioavailable" or "algal available" P (Figure 5-21), quantifying available P in subsoils to determine whether P leaching has occurred, and quick tests that identify the adsorption potential of soils for additional P. Many of these tests would not be conducted routinely, but would be part of a more intensive testing procedure used when routine soil tests identified areas with soil P levels excessive enough to warrant further investigation.

5.5.2 Management of Soil Phosphorus: Fertilization and Conservation Practices

Once the need, or lack of need, of supplemental P has been identified, the next management step is to select the most efficient P source and application technique. From an environmental perspective, the source of P used generally has little influence

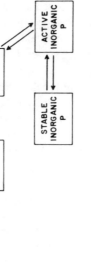
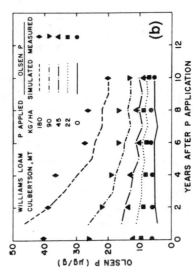

Parameter names and definitions.

C_r = Carbon in the O_r (kg C ha^{-1})
F_{cn} = C/N ratio factor for soil organic matter transformations (unitless)
F_{cp} = C/P ratio factor for soil organic matter transformations (unitless)
F_{il} = Labile P factor for plant P uptake (unitless)
F_{im} = Moisture factor for inorganic P transformation (unitless)
F_{it} = Temperature factor for inorganic P transformations (unitless)
F_l = Fraction of P_l extractable with anion exchange resin after incubation (unitless)
F_{om} = Moisture factor for soil organic matter transformations (unitless)
F_{ot} = Temperature factor for soil organic matter transformations (unitless)
F_{pp} = Plant P stress factor (unitless)
F_{ps} = Plant P scaling factor (unitless)
F_{pu} = Labile P factor affecting R_{pu} (unitless)
F_r = Rooting factor equal to the fraction of roots in a soil layer (unitless)
GS = Plant growth stage (unitless)
K_{as} = Rate constant for the movement of P from P_{ia} to P_{is} (d^{-1})
K_{or} = Rate constant for decomposition of O_r (d^{-1})
K_{os} = Rate constant for decomposition of O_s (d^{-1})
N_i = Inorganic N (NO$_3^-$ + NH$_4^+$) (kg N ha^{-1})
N_{or} = Residue N (kg ha^{-1})
O_m = Organic matter in microbial biomass (kg organic matter ha^{-1})
O_p = Plant organic matter content (kg organic matter ha^{-1})
O_r = Organic matter in crop residue plus microbial biomass (kg organic matter ha^{-1})
O_{rx} = Initial residue organic matter after residue incorporation (kg organic matter ha^{-1})

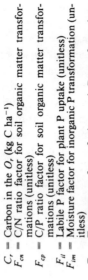
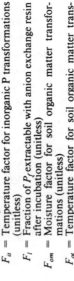
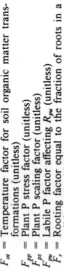
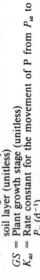

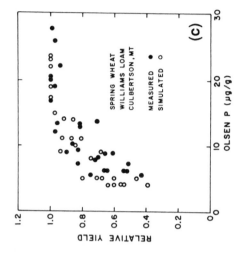

O_s = Stable soil organic matter (kg organic matter ha^{-1})
P_f = Fertilizer P (kg P ha^{-1})
P_i = Inorganic soil P (kg P ha^{-1})
P_{ia} = "Active" inorganic soil P (kg P ha^{-1})
P_{il} = Labile inorganic P (kg P ha^{-1})
P_{ili} = Labile inorganic P prior to fertilization (kg P ha^{-1})
P_{ilc} = Critical level of P_{il} below which R_{pu} is reduced (kg P ha^{-1})
P_{ilf} = Labile P after fertilization and incubation (kg P ha^{-1})
P_{is} = "Stable" inorganic soil P (kg P ha^{-1})
P_m = P in microbial biomass (kg P ha^{-1})
P_o = Organic P in crop residues and stable soil organic matter (kg P ha^{-1})
P_{or} = Organic P in crop residue plus microbial biomass (kg P ha^{-1})
P_{os} = Organic P in stable soil organic matter (kg P ha^{-1})
P_p = Plant P content (kg P ha^{-1})
P_{pg} = Plant grain P pool (kg P ha^{-1})
P_{po} = Optimum plant P content (kg P ha^{-1})
P_{pr} = Plant root P (kg P ha^{-1})
P_{ps} = Plant shoot P (kg P ha^{-1})
R_{as} = Rate of P movement from P_{ia} to P_{is} (kg ha^{-1} d^{-1})
R_{la} = Rate of P movement from P_{il} to P_{ia} (kg ha^{-1} d^{-1})
R_{or} = Rate of O_r decomposition (kg ha^{-1} d^{-1})
R_p = Net mineralization or immobilization from all sources (kg ha^{-1} d^{-1})
R_{pdm} = Rate of plant organic matter increase (kg ha^{-1} d^{-1})
R_{pos} = Rate of P mineralization from P_{os} (kg ha^{-1} d^{-1})
R_{pr} = Rate of gross mineralization of P from decaying O_r (kg ha^{-1} d^{-1})
R_{pu} = Rate of plant uptake of P (kg ha^{-1} d^{-1})
R_{upr} = Gross immobilization (uptake) of P by decomposing O_r (kg ha^{-1} d^{-1})
T = Soil temperature (°C)
W = Volumetric soil water (cm cm^{-1})
$W_{0.03}$ = Volumetric soil water at -0.03 MPa matric potential (cm cm^{-1})

Figure 5-16 Illustration of the parameters required and the use of the EPIC model for P cycling in soils. Figures represent: (a) the major components of the model and simulations conducted using the model to (b) predict changes in soil test P with time and (c) crop yield response to soil test P. (Adapted from Jones et al., 1984a and 1984b. With permission.)

SITE CHARACTERISTIC *(Weighting Value)*	PHOSPHORUS LOSS RATING (VALUE)				
	NONE (0)	LOW (1)	MEDIUM (2)	HIGH (4)	VERY HIGH (8)
Soil Erosion *(1.5)*	N/A	< 11 Mt/ha	11-22 Mt/ha	23-34 Mt/ha	> 34 Mt/ha
Irrigation Erosion *(1.5)*	N/A	QS < 6 for very eerodible soils QS < 10 for other soils	QS > 10 for erosion resistant soils	QS > 6 for erodible soils	QS > 6 for very erodible soils
Soil Runoff Class *(0.5)*	N/A	Very Low or Low	Medium	High	Very High
Soil Test P *(1.0)*	N/A	Low	Medium	High	Excessive
P Fertilizer Rate (kg P/ha) *(0.75)*	None Applied	< 15	16-43	44-73	> 73
P Fertilizer Application Method *(0.5)*	None Applied	Placed with planter deeper than 5 cm	Incorporate immediately before crop	Incorporate > 3 months before crop or surface applied > 3 months before crop	Surface applied > 3 months before crop
Organic P Source Application Rate (kg P/ha) *(1.0)*	None Applied	< 15	16-30	31-43	> 43
Organic P Source Application Method *(1.0)*	None	Injected deeper than 5 cm	Incorporate immediately before crop	Incorporate > 3 months before crop or surface applied < 3 months before crop	Surface applied to pasture or > 3 months before crop

PHOSPHORUS INDEX FOR SITE (TOTAL OF WEIGHTED RATINGS)	SITE VULNERABILITY
< 8	LOW
8 - 14	MEDIUM
15 - 32	HIGH
> 32	VERY HIGH

Figure 5-17 A proposed phosphorus index system to rate sites for their potential to deliver P to sensitive water bodies. (Adapted from Lemunyon and Gilbert, 1993.)

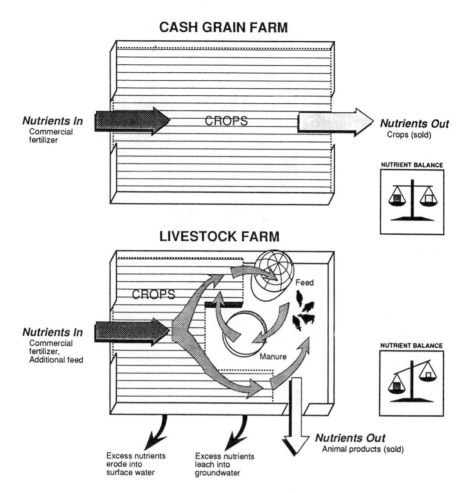

Figure 5-18 Comparison of the nutrient balances for a cash grain and a livestock farm, illustrating the more complex nature of the nutrient cycle for a livestock operation. (Adapted from Weidner, 1988.)

on P loss via runoff or leaching. Application techniques and tillage practices, however, can greatly influence the efficiency of crop recovery of P and the potential for P loss in erosion and runoff.

In general, the most efficient means to avoid P loss is to incorporate fertilizers and manures with the soil, where adsorption and precipitation reactions will reduce P solubility and interaction with runoff waters. Fertilizer placement techniques such as subsurface banding near the seed or seedling are generally superior to broadcast applications, because they reduce the distance required for P diffusion to plant roots and the extent of contact of fertilizer P with soil particles. Most incorporation techniques for large amounts of P fertilizers or organic wastes, however, require soil tillage (moldboard plowing, disking), which often results in soil and P loss via erosion. The effects of a variety of conservation tillage practices ("no-till," "minimum till," "chisel plowing," etc.) on P loss from fertilizers and manures have been

Table 5-6   Statewide and Farm Scale P Budgets for Poultry-Based Agriculture in Delaware

Statewide P Budget

Crop	Hectares	Annual P requirement[a,b]	P Source	Annual amount of available P[c]
Corn	69,600	937,000	Poultry manure	3,500,000
Soybeans	80,600	1,084,000	Fertilizer sales	3,000,000
Wheat	24,300	330,000	Other wastes	Undocumented
Barley	11,000	147,000		
Other	32,400	436,000		
Total	217,900	2,934,000	Total	6,500,000+

Statewide Phosphorus Balance: An Estimated Excess of 3,566,000 kg per State per Yr

Farm Scale P Budget

Nutrient	Needed on farm[d]	Available from manure[d]	Potential excess[d]
N	42,000	23,800	No excess
P	14,500	21,900	+7,400
K	17,000	19,000	+2,000

Note: Assumptions: Crop acreage: 410 ha (corn, soybeans, vegetables); soil tests: 60% of fields test high in P and K; manure amount: 149,000 chickens = 950 Mg/yr; manure analysis: 2.5% N, 2.3% P_2O_5, 2.0% K_2O; manure rate: applied based on crop N requirement.
[a]   Based on current fertilizer recommendations and soil test summaries indicating 65% of the cropland in the state is very high in P.
[b]   kg/Crop.
[c]   kg/Source.
[d]   kg/Farm.

investigated in recent years. As shown in Table 5-4, the presence of crop residues on the soil surface in conservation tillage systems can appreciably reduce the loss of P. Other erosion control practices, such as contour plowing, terracing, buffer strips, and grassed waterways can also reduce erosion, runoff, and thus P loss. However, in many conservation tillage situations, unless fertilizer P is banded or manure P is injected, these P sources are placed directly on the soil surface without incorporation. In some situations, therefore, a cultural operation — conservation tillage — that is designed to control one environmental issue (soil erosion, sediment-bound P loss) may create another (soluble P loss) if P levels in the upper few centimeters of the soil surface become excessive due to long-term overfertilization or manuring. Nutrient management programs that maintain soil P at reasonable levels, or reduced tillage practices that partially incorporate fertilizers and manures (e.g., chisel plowing, Figure 5-22) can help avoid the problem of excessive soluble P loss in runoff from soils.

Conservation tillage practices are particularly difficult to reconcile with animal waste applications where equipment to incorporate manures without tillage is frequently not available. Applying solid or liquid manures directly to the surface of untilled soils not only produces a stratification of P in the topsoil horizon with very high levels present in the erosion prone upper 3 cm, but also leaves P-enriched, highly erodible organic matter on the surface. Complicating the matter further is the

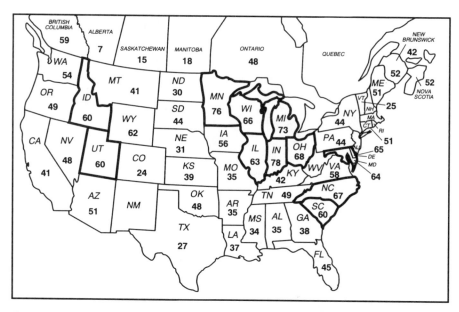

Figure 5-19 Soil test summaries for P in the U.S., 1989. Numbers in each state are percentage of samples testing high or excessive. Highlighted states have 60% or greater of soil test samples testing in the high or excessive range. (Adapted from PPI, 1989.)

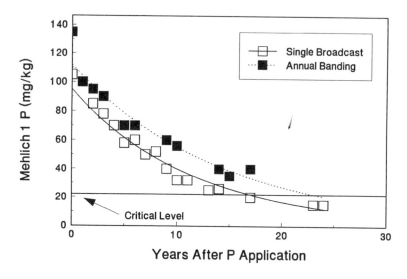

Figure 5-20 Decrease in soil test (Mehlich 1, 0.05 N HCl + 0.025 N H$_2$SO$_4$) extractable P in a Portsmouth soil cropped to corn-soybean rotation for 26 years. Initial soil test P levels (T = 0 years) were established by a single P broadcast application of 324 kg/ha or eight annual banded applications of P at 60 kg/ha. (Adapted from McCollum, 1991.)

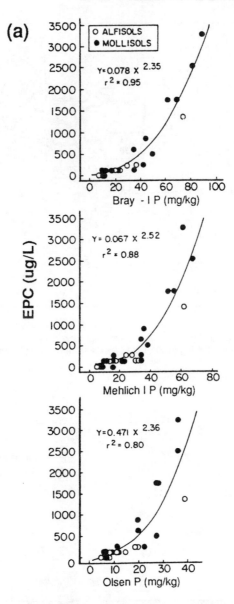

Figure 5-21 Examples of the use of routine soil tests (Olsen P, Mehlich-1 P, Bray-1 P) to predict parameters useful for the assessment of the environmental impact of soil P: (a) EPC_0, and (b) algal-available P. (Adapted from Wolf et al., 1985. With permission.)

fact that manure applications (and P fertilization) are commonly done after the growing season, when farmers have more time available or when frozen ground makes it easier to move equipment through the fields. Fall, winter, and early spring surface applications of inorganic or organic sources of P greatly enhance the likelihood of P losses during the spring when rainfall and snowmelt normally result in the

(b)

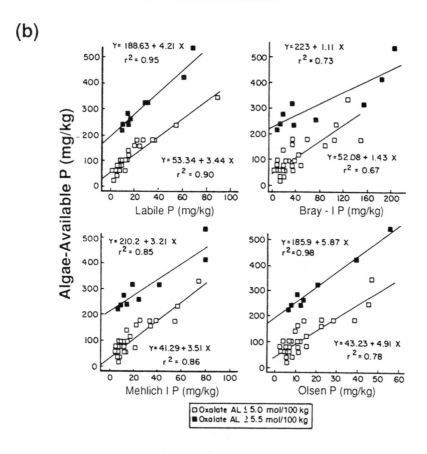

Figure 5-21 (continued)

greatest percentage of annual rainfall occurring. Similar situations occur in agricultural and urban soils that are rarely plowed, such as pastures and turf. In these situations, where particulate P loss is low, the best management strategy that can be used is to avoid unnecessary applications of P that will result in high soluble P losses.

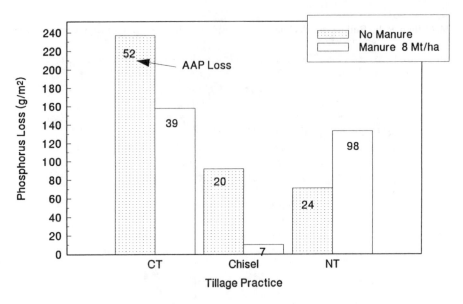

Figure 5-22 Effects of tillage practices and manure use on the loss of total P and algal-available P (AAP = values inside bars) in runoff from corn fields (CT = conventional, Chisel = chisel plow, NT = no tillage). (Adapted from Mueller, et al., 1984.)

REFERENCES

Adriano, D. C., Novak, L. T., Erickson, A. E., Wolcott, A. R., and Ellis, B. G., Effect of long term land disposal by spray irrigation of food processing wastes on some chemical properties of the soil and subsurface water, *J. Environ. Qual.,* 4, 242, 1975.

Alberts, E. E., Neibling, W. E., and Moldenhauer, W. C., Transport of sediment nitrogen and phosphorus in runoff through cornstalk residue strips, *Soil Sci. Soc. Am. J.,* 45, 1177, 1981.

Andraski, B. J., Mueller, D. H., and Daniel, T. C., Phosphorus losses in runoff as affected by tillage, *Soil Sci. Soc. Am. J.,* 49, 1523, 1985.

Chang, A. C., Page, A. L., Sutherland, F. H., and Grgurevic, E., Fractionation of phosphorus in sludge-affected soils, *J. Environ. Qual.,* 12, 286, 1983.

Fixen, P. E. and Grove, J. H., Testing soils for phosphorus, in *Soil Testing and Plant Analysis,* Westerman, R. L., Ed., American Society of Agronomy, Madison, WI, 1991, 141.

Foth, H. D. and Ellis, B. G., *Soil Fertility,* John Wiley & Sons, New York, 1988.

Gachon, L., Les methodes d'appreciation de la fertilite phosphorique des sols, *Bull. Assoc. Fr. Etude Sol,* 4, 17, 1969.

Guertal, E. A., Eckert, D. J., Traina, S. J., and Logan, T. J., Differential phosphorus retention in soil profiles under no-till crop production, *Soil Sci. Soc. Am. J.,* 55, 410, 1991.

Hergert, G. W., Bouldin, D. R., Klausner, S. D., and Zwerman, P. J., Phosphorus concentration-water flow interactions in tile effluent from manured land, *J. Environ. Qual.,* 10, 338, 1981.

Jones, C. A., Cole, C. V., Sharpley, A. N., and Williams, J. R., A simplified soil and plant phosphorus model. I. Documentation, *Soil Sci. Soc. Am. J.,* 48, 800, 1984a.

Jones, C. A., Sharpley, A. N., and Williams, J. R., A simplified soil and plant phosphorus model. II. Testing, *Soil Sci. Soc. Am. J.,* 48, 810, 1984b.

Lemunyon, J. L. and Gilbert, R. G., Concept and need for a phosphorus assessment tool, *J. Prod. Agric.,* in press.

McCollum, R. E., Buildup and decline in soil phosphorus: 30-year trends on a typic umprabuult, *Agron. J.,* 83, 77, 1991.

Mozaffari, P. M. and Sims, J. T., Phosphorus availability and sorption in an Atlantic Coastal Plain watershed dominated by animal-based agriculture, *Soil Sci.,* in press.

Mueller, D. H., Wendt, R. C., and Daniel, T. C., Phosphorus losses as affected by tillage and manure application, *Soil Sci. Soc. Am. J.,* 48, 901, 1984.

Payer, F. S. and Weil, R. R., Phosphorus renovation of wastewater by overland flow application, *J. Environ. Qual.,* 16, 391, 1987.

Pierzynski, G. M. and Logan, T. J., Crop, soil, and management effects on phosphorus soil test levels, *J. Prod. Agr.,* in press.

Potash and Phosphate Institute (PPI), Soil test summaries: phosphorus, potassium and pH, *Better Crops,* 74, 16, 1989.

Ryding, S. O., Enell, M., and White, R. E., Swedish agricultural nonpoint source pollution: a summary of research and findings, *Lake Reserv. Manage.,* 6, 207, 217, 1990.

Sanchez, P. A. and Uehara, G., Management considerations for acid soils with high phosphorus fixation capacity, in *The Role of Phosphorus in Agriculture,* Khasawneh, F. E., Ed., American Society of Agronomy, Madison, WI, 1980, 471.

Sharpley, A. N., The selective erosion of plant nutrients in runoff, *Soil Sci. Soc. Am. J.,* 49, 1527, 1985.

Sharpley, A. N. and Halvorson, A., The management of soil phosphorus availability and its transport in agricultural runoff, *Adv. Agron.,* in press.

Sharpley, A. N., Smith, S. J., Stewart, B. A., and Mathers, A. C., Forms of phosphorus in soil receiving cattle feedlot waste, *J. Environ. Qual.,* 13, 211, 1984.

Sims, J. T., Environmental management of phosphorus in agricultural and municipal wastes, in Future Directions for Agricultural Phosphorus Research, Bulletin Y-224, National Fertilizer and Environmental Research Center, Muscle Shoals, AL, 1992, 59.

Sims, unpublished data.

Weidner, K., Murky water, *Pa State Agric.,* Spring/Summer, 2, 19, 1988.

Wolf, A. M., Baker, D. E., Pionke, H. B., and Kunishi, H. M., Soil tests for estimating labile, soluble, and algae-available P in agricultural soils, *J. Environ. Qual.,* 14, 341, 1985.

Young, R. D., Westfall, D. G., and Colliver, G. W., Production, marketing, and use of phosphorus fertilizers, in *Fertilizer Technology and Use,* Engelstead, O. P., Ed., American Society of Agronomy, Madison, WI, 1985.

SUPPLEMENTARY READING

Mason, C. F., *Biology of Freshwater Pollution,* 2nd ed., John Wiley & Sons, New York, 1991.

Sharpley, A. N. and Menzel, R. G., The impact of soil and fertilizer phosphorus on the environment, *Adv. Agron.,* 41, 297, 1987.

Stewart, J. W. B. and Sharpley, A. N., Controls on dynamics of soil and fertilizer phosphorus and sulfur, in *Soil Fertility and Organic Matter as Components of Production,* Follett, R. F., Stewart, J. W. B., and Cole, C. V., Eds., SSSA Spec. Pub. 19, American Society of Agronomy, Madison, WI, 1987, 101.

Syers, J. K. and Curtin, D., Inorganic reactions controlling phosphorus cycling, in *Phosphorus Cycles in Terrestrial and Aquatic Ecosystems,* Tiessen, H., Ed., UNDP, Saskatchewan Institute Pedology, Saskatoon, Canada, 1988, 17.

Taylor, A. W. and Kilmer, V. J., Agricultural phosphorus in the environment, in *The Role of Phosphorus in Agriculture,* Khasawneh, F. E., Sample, E. C., and Kamprath, E. J., Eds., American Society of Agronomy, Madison, WI, 1980, 545.

6   SOIL SULFUR AND
ENVIRONMENTAL QUALITY

6.1 SULFUR AND THE ENVIRONMENT

Sulfur (S) is an essential nutrient that is a required constituent for all biological systems including plants, animals, and humans. Although S deficiencies and toxicities have been reported for both plants and animals, only cases involving plant S deficiencies appear to be of significance because S-deficient areas have been reported throughout the world. Sulfate (SO_4) is the predominant form of soil S that is absorbed by plants but is not the predominant form in the soil, which explains at least partially why certain areas may show plant S deficiencies. Low S parent materials, high rainfall and intense leaching, and extremely weathered soils, along with low S atmospheric deposition, are common conditions that describe S-deficient areas.

Sulfur occurs naturally in several inorganic and organic forms that are part of the biogeochemistry of the global S cycle (see Chapter 9 for discussion on biogeochemical cycles). The major S pools are part of the lithosphere, similar to that of P. However, the global S cycle is much like that of N, although the major N pool is the atmosphere. Most of the S that cycles through the atmosphere is a result of human activities; although the atmospheric S pool is small, the short mean residence time (MRT) of S compounds actually results in more S cycling through the atmosphere than N (i.e., S flux is greater than N flux). In soils, S transformations are controlled primarily by biochemical processes, which also play an important role in the availability of S to plants.

In this chapter we will discuss the role of S in agricultural, forest, mine land, and aquatic ecosystems, with special emphasis on environmental concern, biogeochemistry, inorganic and organic forms, retention, transformations, nutrition, and management.

6.1.1 Importance of Sulfur

Like N, S is an essential element in several amino acids that are part of plant and animal proteins; S is also an important constituent in vitamins and hormones. Sulfur plays an active role in plant structural composition and metabolic processes, and is crucial to animals because S performs several functions in animal structural, metabolic, and regulator processes. Sulfur has often been overlooked as an essential plant nutrient because until recently S deficiencies were not considered to be very common. The importance of S in plant nutrition and the determination of S availability are discussed further in Sections 6.4.1 and 6.4.2.

For animals, S-containing amino acids are integral components of many biochemicals including proteins, enzymes, and some hormones. Organic S-containing biochemicals are important parts of animal bones, tendons, cartilage, skin, and heart valves. The major structural protein in animals is collagen which contains S amino acids. Enzymatic S plays an important role in enzyme activity and function. Sulfur deficiencies in animals may be manifested as decreased or slow weight gains; lethargism; reduced milk, egg, or wool production; several other symptoms; and if S deficiency is prolonged, possibly death.

The form of S supplied to animals is critical since nonruminants are unable to utilize inorganic S compounds. Ruminants, such as sheep and cattle, are capable of metabolizing inorganic forms of S through the activity of rumen microorganisms. A constant supply of S is required for proper growth in plants and animals, but especially for animals since S is not stored in their bodies.

Sulfur toxicity is also a potential problem to plants and animals. For plants, S toxicity may be exhibited through retarded growth, interveinal chlorosis that extends along leaf margins, or premature senescence of leaves. However, except for citrus, most crops appear to be unaffected by high SO_4; and for those plants that do display symptoms, it is generally the associated cation that is detrimental. The toxic effect of gaseous SO_2 and H_2S compounds on plants is discussed in Section 6.4. As for animals, S toxicity is primarily a function of the S species. Ruminants have suffered from S toxicity when $(NH_4)_2SO_4$ or gypsum $(CaSO_4)$ were used as nonprotein N or Ca sources, respectively. Nonruminants suffer from S toxicity when excessive amounts of S-containing amino acids, especially methionine, are consumed in their diet.

6.1.2 Environmental Impact of Sulfur

Some environmental concerns which are related to S are listed in Table 6-1. Acidic deposition is the result of the oxidation of S and N in the atmosphere and will be discussed in Chapter 10. Acid soils and acid mine drainage occur when reduced S compounds and minerals are oxidized, forming sulfuric acid (H_2SO_4). Examples of acid mine drainage are most prevalent at inactive or abandoned mine sites; in 1969, the source of 78% of acid mine drainage was associated with inactive or abandoned coal mines. Degradation of 16,000 km of streams was attributed to acid mine drainage. Geothermal activity can influence the surrounding soil and plant ecosystems by releasing significant amounts of gaseous and soluble S into the atmosphere. Sulfate is highly soluble and is known to accumulate in groundwaters of arid and semiarid environments. A drinking water standard of 250 ppm (mg SO_4/L) was set because SO_4 was found to act as a laxative to some humans and animals. Sulfur toxicity to animals was noted above.

Burning of fossil fuels; petroleum refining; and processing of Cu, Pb, Zn, and Ni have resulted in enormous amounts of S being emitted into the atmosphere every year (Table 6-2). Various estimates have ranged from approximately 20 to over 50% for the amount of S entering the atmosphere that is contributed by the above sources. In

Table 6-1   Potential Environmental Impacts Due to Sulfur

Environmental concern	Description of problem
Acid deposition	Sulfuric acid (H_2SO_4) formation in the atmosphere results in wet deposition (rain, snow, mist, fog) and dry deposition (particulates) that may be detrimental to vegetation, surface waters, buildings and structures, and humans; see Table 4-1 for environmental impacts due to N and Chapter 10 for further discussion on acidic deposition
Acid sulfate soils	Caused by the release of H_2SO_4 into the soils through the oxidation of sulfidic materials which are commonly associated with coastal regions and lignite coal mining operations
Acid mine drainage	Oxidation of reduced forms of S from mining activities produces H_2SO_4 that can impact the soils or spoils and surface waters in the surrounding environment
Geothermal activity	Geysers and other geothermal releases can bring significant amounts of gaseous and soluble S compounds to the earth's surface; vegetation in the surrounding area can be killed by the extremely acidic soils that form; soil pH as low as 0.9 has been reported in Yellowstone National Park, Wyoming
Groundwater contamination	High SO_4 levels can render groundwaters unusable for human and livestock consumption
Sulfur toxicity to animals	Both ruminant and nonruminant animals can suffer from S toxicity; ruminants are susceptible to $(NH_4)_2SO_4$ and gypsum ($CaSO_4$) when these are used as nonprotein N or Ca feed sources, respectively; nonruminants are susceptible to excessive S amino acids, especially methionine, in their diet; livestock are also sensitive to H_2S (Goodrich and Garrett, 1986)

Table 6-2   Estimations of Global S Emissions from Various Industrial Sources

Source	S emitted (Tg/yr)[a]	Source	S emitted (Tg/yr)
Coal	51	Cr smelting	7.5
Lignite (brown coal and peat)	17.5	Cu refining	1.4
		Pb smelting	0.8
Coke (briquettes)	1.3	Zn smelting	0.5
Total coal	**69.8**	Ni smelting	0.8
		Total metallurgical operations	**11.0**
Petroleum refining	3.2		
Motor gasoline	0.4	Sulfuric acid manufacturing	1.3
Kerosene	0.1	Incineration	0.3
Fuel and diesel	1.7	Pulp and paper	0.3
Petroleum coke	0.2	**Total miscellaneous**	**1.9**
Residual fuel oils	18.0		
Total petroleum	**23.6**	**Total globals S emissions**	**106.3**

Source: Krupa and Tabatabai, 1986.
[a] $Tg = 10^{12}$ g.

industrial regions, where most of the S is released, acidic deposition (wet and dry deposition containing high levels of H_2SO_4) has caused several problems that are currently being addressed by many nations throughout the world. Precipitation in affected areas can have a pH below 4, which, compared to rain in many nonindustrial areas that can have pH levels of 6–7, suggests a 100–1000 fold increase in H^+ ions.

Table 6-3   Forms of Inorganic S in the Environment

Mean oxidation state	Category	Compound	Formula
-2	Sulfides	Sulfide ion	S^{2-}
		Bisulfide ion	HS^-
		Hydrogen sulfide	H_2S
		Carbonyl sulfide	COS
-1	Polysulfides	Disulfide	S_2^{2-}
		Pyrite	FeS_2
0	Elemental	Sulfur	S^0
+2[a]	Thiosulfate	Thiosulfate ion	$S_2O_3^{2-}$
		Bithiosulfate ion	$HS_2O_3^-$
		Thiosulfuric acid	$H_2S_2O_3$
+4	Sulfites	Sulfite ion	SO_3^{2-}
		Hydrogen sulfite	HSO_3^-
		Sulfur dioxide	SO_2
+6	Sulfates	Sulfate ion	SO_4^{2-}
		Bisulfate ion	HSO_4^-
		Sulfuric acid	H_2SO_4

Source: Stevenson, 1986.
[a]   Sulfur oxidation states for thiosulfates are -2 and +6 with an average of +2.

The amount of S that falls annually in rainfall varies depending on the natural and anthropogenic sources. Since plants require S (9–39 kg/ha for average crop yields), some regions benefit from the S added in rainfall. In nonindustrial areas, the amount of S may be sufficient to sustain adequate yield. For example, S (kg/ha) in precipitation has been estimated for several urban (nonindustrial) and rural regions of the U.S. to be 1–37 in the southeast, 4–42 in the midwest, and 5–17 in the north central states. Sulfur deposition in urban and rural regions of other countries are comparable to those listed above; 2–16 in Canada, 13–27 in China, and 10–20 kg S/ha in the former Soviet Union.

6.2 GLOBAL SULFUR CYCLE

Sulfur exists in several inorganic and organic chemical forms within the atmosphere, hydrosphere, and soil environments. Soil S exists as aqueous species such as SO_4^{2-}; solid forms that consist of mineral matter and organic constituents of plant, animal and microbial origin; and gaseous species. Oxidation states for S range from -2 to +6 (see Table 6-3 for examples of oxidation states for inorganic and gaseous S species). As S cycles through atmosphere, hydrosphere, and soil ecosystems, several S transformations can occur — due to biochemical and chemical processes — and form S species of different oxidation states. For example, carbonyl sulfide (COS), the most abundant atmospheric S gas, is oxidized to SO_4 or taken up by plants and metabolized. Sulfate entering soils is utilized by plants and microorganisms whereby the SO_4 is reduced in the formation of amino acids, proteins, and other S-containing biochemicals. Decomposition of organic matter releases oxidized and reduced S forms depending on environmental conditions, in which oxygen plays a dominant role. Several pathways have been proposed for the transformation of organic S to volatile S compounds including the following:

$$SO_4^{2-} \longrightarrow \text{Cysteine} \longrightarrow \text{Methionine} \longrightarrow \text{Volatile S compounds} \qquad (6\text{-}1)$$

Physical processes such as movement of S as dusts, wet and dry deposition, or precipitation-dissolution reactions involving S evaporites are also responsible for the transfer of S from one ecosystem to another.

The largest S pool (or reservoir) is that of the lithosphere, followed by the hydrosphere, and finally the atmosphere which contains only a minor amount of S (Table 6-4). By far the greatest S pool exists in sedimentary pyrite, shale, and evaporite materials. In soils, S concentrations range from a low of approximately 0.002% in course-textured, highly weathered, leached soils in humid areas to greater than 5% in gypsiferous soils of the arid and semiarid areas, and in soils developed in tidal marshes. Soil S is predominately in inorganic and organic forms, and the ratio of the two varies with soil properties (pH, moisture status, organic matter and clay contents), soil depth, and climatic conditions.

In the S cycle, transfer of S from one pool (or reservoir) to another occurs at some rate proportional to the amount entering and exiting a pool and is related to the S flux (Figure 6-1). Although only small amounts of S are transferred from one S pool to another, humans have had an immense influence on the annual atmospheric S flux. As with the increase in C-based materials entering the atmosphere since the beginning of the industrial era (see Chapter 10), so too are S-based materials increasing.

Table 6-4 Estimated Mass of S in Major S Pools

Pool	Mass (kg S)
Atmosphere	3.6×10^9
Hydrosphere	1.3×10^{18}
Oceans	1.3×10^{18}
Marine organisms	2.4×10^{10}
Fresh waters	3.0×10^{12}
Ice	6.0×10^{12}
Lithosphere	24.1×10^{18}
Igneous rocks	5.0×10^{18}
Metamorphic rocks	11.4×10^{18}
Sedimentary rocks	7.7×10^{18}
Evaporites	5.1×10^{18}
Shales	2.0×10^{18}
Limestones	0.1×10^{18}
Sandstones	0.3×10^{18}
Soils	2.6×10^{14}
Soil organic matter	1.0×10^{13}
Biosphere	7.6×10^{12}

Source: Freney, 1986 and Trudinger, 1986.

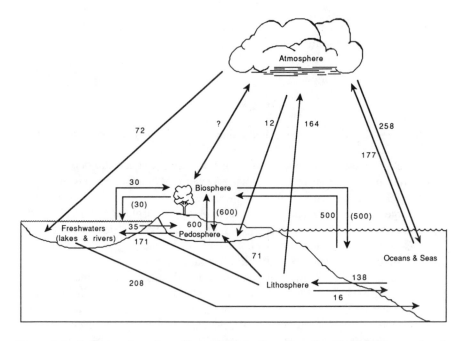

Figure 6-1   Sulfur transformations (fluxes) between the various S pools. Numbers are given in Tg/year (Tg = 10^{12} g) and were taken from Trudinger, 1986. A S balance (inputs = outputs) is presumed between biosphere and atmosphere.

Over the past 100 years, the rate of S entering the atmosphere by fossil fuel emissions has doubled, and currently is approximately 1.5×10^{11} kg S per year. Over time, the amount of oxidized S is increasing, some of which is returned to the earth's surface as wet and dry deposition (i.e., acidic deposition). Further discussion on atmospheric deposition of S and N can be found in Chapter 10.

The soil S cycle is similar in many respects to N and P soil cycles in that each is affected by biochemical and chemical processes. The S cycle — emphasizing soil, plant, and animal S transformations — is shown in Figure 6-2. Four dominant S forms are present in soils and include elemental S, sulfides, sulfates, and organic S compounds. Additions of soluble S to soil results from mineral weathering; atmospheric deposition; fertilization (and other soil amendments); and decomposition of plant, animal, and organic matter. Losses of S occur during leaching, surface runoff, and crop removal. Transformations that take place between the various S forms are mediated primarily by biochemical processes which will be discussed in more detail later in this chapter.

6.2.1 Inorganic Sulfur in Soils

Several forms of inorganic S exist in soils (Table 6-3). In well-drained soils, inorganic S occurs primarily as SO_4 either in a dissolved, adsorbed, or solid state. Soils containing Al and Fe hydrous oxides are capable of adsorbing SO_4 and preventing it from being leached. The S solid form, gypsum ($CaSO_4 \cdot 2H_2O$), is found primarily in calcareous soils of arid and semiarid regions; $CaSO_4$ may also occur as

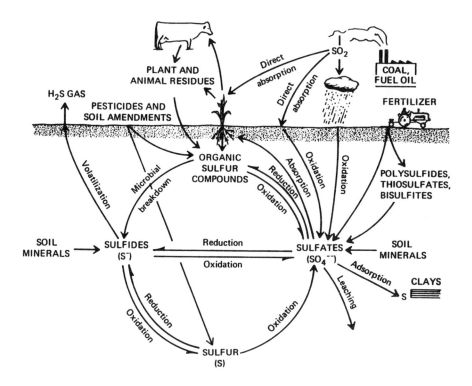

Figure 6-2 The S cycle depicting the various additions, losses, and transformations that occur in soils. (From Brady, 1974. With permission.)

a coprecipitated mineral form associated with $CaCO_3$ (limestone). Other less common solid mineral forms of SO_4 include barite ($BaSO_4$), $SrSO_4$ jarosite [$KFe_3(OH)_6(SO_4)_2$], and coquinbite [$Fe_2(SO_4)_3 \cdot 5H_2O$]. In anaerobic soils such as wetlands, tidal marshes, and poorly drained soils, reduced forms of S are common and can include FeS, FeS_2, H_2S, and $S°$.

Gaseous forms of S are also present in soils and range from minute concentrations in well-drained, aerobic soils to high concentrations in waterlogged, anaerobic soils. Soil S is less volatile than soil C and N compounds, whereas P has essentially no gaseous forms found in nature. Gases formed during organic matter decay which may be emitted from soils include carbon bisulfide (CS_2), COS, methyl mercaptan (CH_3SH), dimethyl sulfide [$(CH_3)_2S$], dimethyl disulfide [$(CH_3)_2S_2$], and hydrogen sulfide (H_2S), all of which have characteristic distinctive odors. Sulfur emissions from anaerobic soils and wetlands (i.e., swamps) are approximately 30×10^{12} g/year.

Elemental S and sulfides are oxidized to H_2SO_4 when exposed to aerobic conditions. Oxidation of reduced S occurs when subsurface geologic materials are brought to the earth's surface during mining activities or on draining a tidal marsh. Under conditions such as these the solution pH level may be sustained as low as 2 or less until the reduced S is oxidized, neutralized, or otherwise removed from the soil environment.

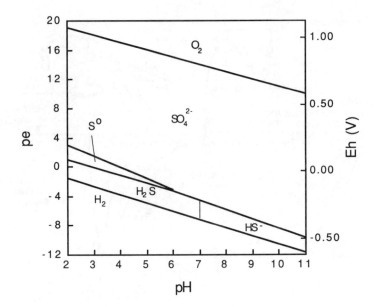

Figure 6-3   Relationship among the S species at different pe and pH values.

Speciation of inorganic S is controlled by pe (e⁻ activity or oxidation/reduction condition) and pH (H^+ activity) relationships as shown in Figure 6-3. Sulfate is the dominate species found at high pe levels and all levels of pH. At low to intermediate pe levels, sulfide species predominate with H_2S at low pH (<7) and HS^- at high pH (>7) levels. Elemental S can also exist at moderate pe levels and acid pH. Some S-containing amino acids are also stable at pe + pH levels within the SO_4^{2-} range, thus suggesting organic S compounds may persist in aerobic soil environments.

6.2.2 Organic Sulfur in Soils

Organic S is the dominant form of S in most soil and aquatic environments. The distribution of S in several types of soils is listed in Table 6-5, which clearly shows that organic S is the predominant form of soil S. Only in low organic matter, calcareous soils typical of arid and semiarid regions will concentrations of inorganic S eclipse organic S. Further examination of the inorganic and organic S compounds can be obtained by S fractionation methods as discussed in the next section.

The content of organic S in soil and aquatic environments varies considerably, depending primarily on the amount of organic matter present. In soils, organic matter accumulates at the soil surface as plant materials decay; subsurface horizons can also accumulate organic S as shown in Table 6-6. Note in Table 6-6 that total and organic S in soil 1 decreases with depth while inorganic S stays relatively constant, whereas in soil 2 there is an accumulation of all three forms of S that occurs at the 30–60 cm depth. Soil 2 is typical of Spodosol (Podzol) type soils, which through translocation of organic matter into subsurface horizons have accumulated organic S. The increase in inorganic S is related to increased SO_4 adsorption by Fe and Al hydrous oxides that also accumulated in the subsurface horizons of Spodosols.

Table 6-5   Distribution of Inorganic and Organic S in Different Soils

Type of soil or land use	Location	Total S (mg/kg)	Inorganic S (%)	Organic S (%)
			Range	
Agricultural	Iowa[a]	78–452	1–3	97–99
	Brazil	43–398	5–23	77–95
	West Indies	110–510	2–10	90–98
	Alberta[b]	80–700	8–15	85–92
	Queensland[c]	11–725	2–18	82–98
	New South Wales[c]	38–545	4–13	87–96
	New Zealand	240–1360	2–9	91–98
Forest[d]	New Hampshire[a]	452–1563	1–8	92–99
	New York[a]	68–2003	1–18	82–99
	Alberta[b]	364–1593	2–10	90–98
	Illinois[a]	112–555	2–10	90–98
	Germany	74–328	7–28	72–93
Surface	Iowa[a]	55–618	1–5	95–99
Organic	England	7405	5	95
Acid	Scotland	300–800	2–10	90–98
Calcareous	Scotland	460–1790	21–89	11–79
Volcanic ash	Hawaii[a]	180–2200	6–50	50–94

Source: Freney, 1986 and Mitchell et al., 1992.
[a] United States.
[b] Canada.
[c] Australia.
[d] Range values represent variation in a single profile.

Organic S compounds in soils and aquatic environments are derived from several plant, animal, and microbial sources. Sulfur-containing amino acids are generally the most prevalent of the known organic S compounds, accounting for 10–30% of the total S in various soils. These amino acids are readily decomposed and may only exist for short periods of time before they are utilized by microorganisms. In addition, S-containing amino acids are often bound to mineral and organic matter, making their extraction from soils difficult. Sulfur-containing polysaccharides and lipids are also present in soils and may comprise approximately 5% of the total S in some soils.

Sewage sludge and manures contain several forms of organic S. Although these materials are perceived as waste products that require disposal, several benefits can be realized when utilizing sewage sludge and manures in a land application program.

Table 6-6   Relationship of Total, Inorganic, and Organic S by Depth in Two Soil Profiles

Depth (cm)	Total S		Inorganic S		Organic S	
	Soil 1	Soil 2	Soil 1	Soil 2	Soil 1	Soil 2
0–10	436	205	7	8	429	197
10–20	389	93	6	5	383	88
20–30	342	118	6	7	336	111
30–45	266	134	6	18	260	116
45–60	230	134	8	29	222	105
60–75	188	115	7	22	181	93
75–90	152	99	5	10	147	89

Source: Williams, 1974.
Note: Values given in mg S/kg soil.

In addition to the organic matter, N, and P contained in sludges and manures, both inorganic and organic forms of S are generally part of these materials. For example, some sewage sludges have been found to contain elemental S and sulfides; and organic S such as S-containing amino acids, alkyl benzene sulfonates, sulfonated acidic polysaccharides, and ester sulfates (a significant portion of which is related to detergents).

6.2.3 Sulfur Fractionation Scheme

Due to the heterogeneous nature of organic matter (see Chapter 2), soil and aquatic organic S compounds are generally determined by fractionation methods. The S components identified in a S fractionation method are illustrated in Figure 6-4, which indicates S can be separated into five basic fractions plus total S. Both inorganic S (SO_4 and nonsulfate S, primarily sulfides) and organic S (C-bonded and ester sulfate) that have been mentioned in the previous sections can be determined by this method. Examples of concentrations and proportions of the S fractions in sewage sludge, soils, and lake sediments and waters are listed in Table 6-7. The S fractionation method can also be used to evaluate soil, aquatic, and possibly atmospheric S dynamics including transformations and flux rates.

The nature and content of soil S can be quite variable. Using the S fractionation method, total S is determined on a subsample of the soil. Water soluble and phosphate extractable S are also determined on subsamples, and the difference between the two is considered specifically adsorbed SO_4 (SO_4 adsorption will be discussed in the next section). Inorganic nonsulfate S is evaluated directly by Zn-HCl digestion. Organic S is comprised of both ester sulfate (–C–O–S–) and C-bonded (–C–S–) forms of which the former is determined along with inorganic S through the hydriodic acid (HI) digestion. C-bonded S is obtained by difference (total S minus HI) since there is no method available that can accurately and completely account for all the C-bonded S in a sample.

The distribution of inorganic and organic S in several soils was shown in Table 6-5. Organic S, which consists of ester sulfate (Figure 6-5a) and C-bonded S (Figure 6-5b), is the dominant S form found in most soils. There appears to be a difference in the content of organic S forms in agricultural vs forested soils. The percentage of total S in agricultural soils that is comprised of ester sulfates ranges between 35–60% whereas in forested soils the range is lower at 20–30%. The amount of C-bonded S in forested soils typically averages 50–80% of the total S. In agricultural soils, lower percentages of ester sulfates have been attributed to the removal of C-bonded S with cropping.

6.3 SULFUR RETENTION AND TRANSFORMATIONS IN SOILS

There are many processes that occur in soil and aquatic environments that affect S movement and its transformation, but the most important processes are adsorption-desorption of SO_4, inorganic S oxidation-reduction, and organic S transformations. It has been pointed out in previous sections that organic S comprises the greatest

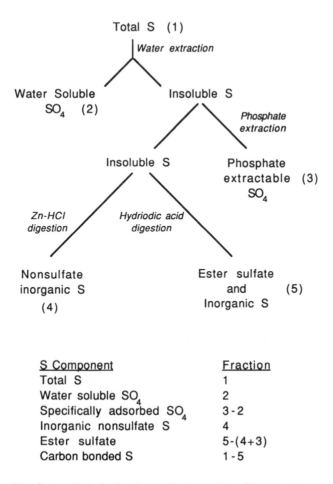

S Component	Fraction
Total S	1
Water soluble SO_4	2
Specifically adsorbed SO_4	3 - 2
Inorganic nonsulfate S	4
Ester sulfate	5 - (4 + 3)
Carbon bonded S	1 - 5

Figure 6-4 Schematic indicating the methods used in a S fractionation procedure.

proportion of total S in most soils. Organic S transformations, therefore, are of particular importance when considering soil S cycling and reactions.

6.3.1 Adsorption and Desorption of Soil Sulfate

Adsorption of SO_4 by soils results in the retention of S that could otherwise have been leached from the soil profile. Since SO_4 is the major form of S that is taken up by plants, the retention of SO_4 may enhance the ability of the soils to supply S to plants in the future. The amount of water soluble and exchangeable SO_4 can be low in some soils, which is dependent on factors such as soil type, climate, and management practices. The mineralogical composition of a soil can have a significant impact on its ability to retain adsorbed SO_4. For example, a Tennessee forest soil had an adsorbed SO_4 pool that was 15 times greater than the total S stored in the aboveground biomass.

Sites for SO_4 adsorption include Fe and Al hydrous oxides, edges of aluminosilicate clays, and possibly metal-organic matter complexes. In some soils, coatings of

Table 6-7   Average S Constituents[a] in Solid or Aqueous Samples from
Different Ecosystems

Sample	Total S	Extractable SO₄	Inorganic nonsulfate S	Ester S	Carbon bonded-S
Sewage sludge	344	37 (11)	46.2 (13)	126 (37)	134 (39)
Soil (Oi horizon)	50.0	0.48 (1)	0.62 (1)	7.16 (14)	41.9 (84)
Soil (Bhs horizon)	16.5	0.72 (4)	0.69 (4)	3.59 (22)	11.5 (70)
Lake sediment	229	35 (15)	1.9 (1)	45.3 (20)	146 (64)
Lake water[b]	0.064	0.053 (83)	0.00 (0)	0.007 (11)	0.004 (6)

Source: Landers et al., 1983.
Note: Values in parentheses are the percent of total S in that S fraction.
[a] μmol/g.
[b] Values reported as μmol/L.

Fe and Al hydrous oxides on soil particles (i.e., spodic B horizons) contribute to the majority of SO_4 adsorption sites. Sulfate adsorption is expected to be greater on amorphous or poorly crystalline Fe and Al hydrous oxides because they have greater surface area than crystalline forms. The number of edge sites on aluminosilicate clays determines the amount of SO_4 adsorbed. Some studies have shown that 1:1 clays adsorb greater amounts of SO_4 than do 2:1 clays. Sulfate adsorption by metal-organic complexes is proposed to occur by forming SO_4-metal-organic complexes.

Sulfate adsorption is influenced by factors such as pH, amount and crystallinity of Fe and Al hydrous oxides, and presence of organic matter. Maximum adsorption of SO_4 occurs at low pH, typically around pH 3–4.5, which is due to a net positive surface charge that develops on Fe and Al hydrous oxides. The increasing number of positively charged sites with decreasing pH enhances SO_4 adsorption; however, Fe and Al hydrous oxides are not stable at extremely low pH, and their dissolution will reduce the number of adsorption sites available. Organic matter can effectively hinder SO_4 adsorption by coating the mineral and hydrous oxide surfaces or competing with SO_4 for the adsorption sites.

Desorption of SO_4 may occur in soils if the soils have received high S inputs and have a relatively low SO_4 adsorption capacity, and/or SO_4 is displaced by competing anions. The amount of readily desorbable SO_4 decreases with time, which has been attributed to the formation of more tightly bound (specifically adsorbed) SO_4 or a conversion of the SO_4 to organic forms. Desorption of SO_4 is enhanced by using a phosphate extracting solution which will readily displace SO_4 that is present in specifically bound forms.

Sulfate adsorption can be analyzed by using Langmuir and Freundlich mathematical models (see Chapter 5 for a discussion of these models in P adsorption studies), and a recently proposed model called the initial mass (IM) isotherm. The IM isotherm was shown to fit forest soil SO_4 adsorption data better than Langmuir and Freundlich models which were unable to (1) provide a linear relationship, or (2) be transformed for use in the linearization process. The IM isotherm is written as:

$$SO_4 \text{(adsorbed or released)} = mX_i - b \qquad (6\text{-}2)$$

(a) Ester Sulfates

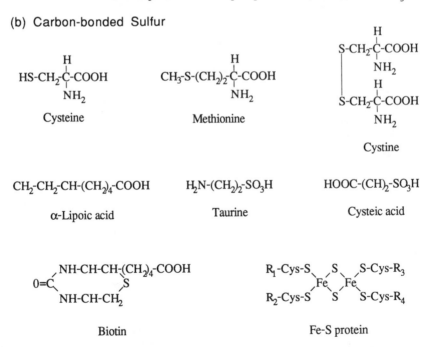

*Adenosine-5'-phosphosulfate (APS), if X = H

3'-Phosphoadenosine-5'-phosphosulfate (PAPS), if X = PO_3^{2-}

(b) Carbon-bonded Sulfur

Figure 6-5 Various organic S compounds that are present in soils. Represented are (a) ester sulfate compounds and (b) carbon-bonded compounds.

where

m = slope
X_i = initial amount of solution SO_4 with respect to the soil mass
b = intercept

Since native SO_4 levels are high in some soils, there are cases where SO_4 release occurs. The SO_4 reserve soil pool (RSP) can be calculated from the IM isotherm as:

$$RSP = \frac{b}{1-m} \qquad (6\text{-}3)$$

The RSP is a measure of the labile soil SO_4, and is an estimate of biologically active SO_4 in a soil.

6.3.2 Inorganic Sulfur Oxidation-Reduction Reactions

Oxidation-reduction (redox) reactions involving inorganic S are primarily mediated by microbial processes. The role of S redox reactions in soils were shown in Figure 6-2, which portrays several pathways available for the oxidation or reduction of S. As mentioned earlier, SO_4 is the dominant inorganic form of S in aerobic soils. Sulfides are more important in anaerobic soils such as found in wetlands, swamps, and tidal marshes. Therefore, reduction reactions involving SO_4 and oxidation reactions involving S^{2-} are considered here.

Sulfate reduction occurs by assimilatory or dissimilatory reduction. The former reduction process occurs when microorganisms utilize SO_4 in the assimilation of cellular constituents. In converting inorganic S to organic S, the S is said to be immobilized. The dissimilatory reduction process is unique to a specific genera of bacteria, of which *Desulfovibrio* is the most well-known, that reduce SO_4 to sulfides. Dissimilatory SO_4 reduction occurs in anaerobic environments such as in water-logged soils, swamps, stagnant waters, and tidal marshes.

Conditions under which dissimilatory S-reducing microorganisms are capable of surviving are quite varied. They can live in soils having an extensive pH range and high salt content such as in saline lakes, evaporation beds, deep-sea sediments, and oil wells. A low redox potential (E_h) of less than 100 mV (pe = 1.69) is required for their growth; under higher redox potentials they can subsist in a dormant state.

Environmental problems associated with SO_4 reduction are varied. In sewage systems, the production of H_2S leads to excessive corrosion of stone, concrete, and pumps and other components of the sewage distribution system. Hydrogen sulfide also has an odor similar to that of rotten eggs. The presence of reduced forms of S in fossil fuels, sediments, and anaerobic soils can cause a number of problems as the S oxidizes back to SO_4, which will be discussed next.

Oxidation of reduced inorganic S compounds occurs under aerobic conditions and follows several pathways. Although S oxidation is considered to be primarily biochemically driven, chemical oxidation of sulfides (S_2^-), sulfites (SO_3^{2-}), and thiosulfates occurs readily in nature; however, the rate of chemical oxidation is slower than that of the biochemical oxidation processes. Oxidation of reduced S is also an acidifying process, which is a cause for concern in certain situations. Examples of S oxidation that result in the formation of acidity are:

$$H_2S + 2O_2 \rightleftharpoons H_2SO_4 \rightleftharpoons 2H^+ + SO_4^{2-} \qquad (6\text{-}4)$$

$$2S + 3O_2 + 2H_2O \rightleftharpoons 2H_2SO_4 \rightleftharpoons 4H^+ + 2SO_4^{2-} \qquad (6\text{-}5)$$

There are several types of autotrophic and heterotrophic microorganisms capable of oxidizing S. Bacteria of the *Thiobacillus* genus are capable of surviving over a broad range of soil and environmental conditions; some *Thiobacillus* species are capable of living in soil ranging in pH from 1.5 to 9. *Sulfolobus* are another group of S-oxidizing organisms that live in S-rich geothermal hot springs. They survive in waters that range in pH from 2 to 5 and temperatures of 60–80°C. Another form of S-oxidizing microorganisms that survives in hot springs is the *Thermothrix;* but unlike *Sulfolobus, Thermothrix* grows in near neutral pH waters.

In neutral and alkaline pH soils, the primary S oxidizers may be a group of heterotrophic microorganisms. The heterotrophic microorganisms of interest include several genera of bacteria (*Arthrobacter, Bacillus, Micrococcus, Mycobacterium,* and *Pseudomonas*), as well as some actinomycetes and fungi which are capable of oxidizing inorganic S compounds. These microorganisms are believed to oxidize inorganic S only as a consequence of other metabolic processes.

Factors that influence S oxidation in soils include soil type, pH, temperature, moisture, and organic matter. Oxidation of inorganic reduced forms of S is an acidifying process that can develop acidic soils that are toxic to plants, animals, and microorganisms. Acid mine drainage occurs when reduced forms of S are oxidized during the mining process. Oxidation of pyrite, a common S-containing mineral found in reduced subsurface environments, occurs as follows:

$$2FeS_2 + 7H_2O + 7.5O_2 \rightleftharpoons 8H^+ + 4SO_4^{2-} + 2Fe(OH)_3 \qquad (6\text{-}6)$$

6.3.3 Organic Sulfur Transformations

The transformation of organic S to inorganic S and volatile gases are processes mediated by the activity of soil microorganisms. Release of inorganic S during the decomposition of organic matter is important for supplying adequate S for plant growth. Soil factors that influence the growth and activity of microorganisms (i.e., pH, temperature, soil moisture, organic matter) will also affect the rate of organic S transformations.

Transformation of organic S to inorganic S, and of inorganic S to organic S are processes that describe S mineralization and immobilization, respectively, as shown in the following equation:

$$\text{Organic S} \; \underset{\text{Immobilization}}{\overset{\text{Mineralization}}{\rightleftharpoons}} \; \text{Inorganic S} \qquad (6\text{-}7)$$

During the mineralization of S, inorganic S, — primarily SO_4, — is released as microorganisms decompose organic matter, utilizing some of the S for synthesizing cell constituents and releasing some inorganic S to the soil solution. Immobilization of S, on the other hand, is a process in which inorganic S is assimilated by micro-organisms when low S organic energy sources (i.e., organic matter) are added to soils. Mineralization-immobilization processes were also discussed for N and P in earlier chapters.

6.4 PLANT RESPONSE TO SULFUR

Sulfur is one of the secondary plant nutrients (S, Ca, Mg), but it has been considered by many as the fourth most important plant nutrient after N, P, and K. Sulfur deficiencies have become increasingly common because fertilizers which used to contain S (e.g., normal superphosphate, $[(NH_4)_2SO_4]$) have been replaced by low-S fertilizers (triple superphosphate, $[(NH_4)_2HPO_4]$); S-containing pesticides have been converted to organic-based pesticides, and atmospheric S inputs have been reduced — all of which have lowered the amount of S added to soils. Since the development of new high grade fertilizers with low S impurities, S deficiencies have become more evident in many parts of the world. Additional demands for higher crop yields have also put a greater demand on soil S reservoirs to supply extra amounts of available S for increased plant growth. Reports of S deficiencies and increased crop yields due to S fertilization have been reported in several regions throughout the U.S. and in other parts of the world.

Gaseous atmospheric S compounds may have a detrimental, beneficial, or neutral effect on plants. The concentrations of ambient gaseous S forms, SO_2 and H_2S, are low and can be beneficial; however, in the vicinity of high discharge areas, the concentration of these ambient gases may be harmful to plants. Plants absorb SO_2 primarily through the stomata and are capable of oxidizing the SO_2 to sulfites and sulfates. However, when plants are exposed to levels of SO_2 that exceed the plant's ability to oxidize sulfite to sulfate, sulfite can accumulate in the plant to toxic levels. Visual symptoms show up on affected leaves that can be mistaken for other types of injury; noting the close proximity of a SO_2-generating source will help in diagnosing whether S toxicity has occurred. Short-term exposure of plants to H_2S generally causes little, if any damage; whereas continuous exposure to low levels, such as near geothermal sites, may affect plants by causing leaf lesions, defoliation, and decreased growth.

6.4.1 Sulfur in Plant Nutrition

Sulfur requirements vary with plant species, cultivars, and stages of development. Several S-containing compounds are present in plants, but the majority of the S is part of the S amino acids cysteine, cystine, and methionine, which are found largely in proteins. Additional S-containing compounds include vitamins, biotin, thiamine, and B_1; and other coenzymes, lipoic acid and coenzyme A (CoA). Sulfur deficiencies not

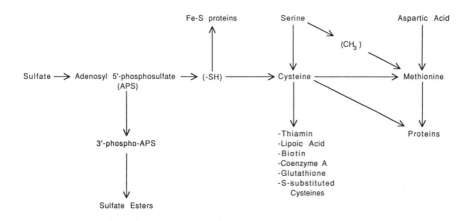

Figure 6-6   Pathways involving the transformation of SO_4 through intermediates to different S-containing organic compounds in plants. The reduction of SO_4 in APS is denoted -SH. (Adapted from Thompson et al., 1986.)

only may reduce yields due to improper plant nutrition, but also may lower the quality of certain crop products (e.g., digestibility of forages and baking quality of flours).

Figure 6-6 illustrates the various S pathways that are involved in the synthesis of organic S compounds, starting with SO_4. Once S (primarily SO_4) enters the plant, it is activated through the formation of adenosine-5'-phosphate (APS). APS is a precursor to 3'-phospho-APS (PAPS), which is an intermediate in the formation of ester sulfates. The activated SO_4 in APS is reduced, and the reduced S is incorporated into either Fe-S proteins or cysteine which acts as a precursor to several organic S compounds (thiamin, lipoic acid, biotin, coenzyme A, glutathione, and S-substituted cysteines). The majority of the two S-containing amino acids, cysteine and methionine, are incorporated into various proteins.

A number of important plant functions are the result of S-containing compounds; in fact, plant metabolism and physiology are dependent on these compounds. Methionine and cysteine are essential components to many plant enzymes; and their structure and function are, in part, due to S interactions. The Fe-S protein — ferredoxin — is involved in oxidation-reduction reactions, most notably as a stable redox compound in the photosynthesis process. Nitrogen metabolism involving nitrate reduction, N_2 fixation, and NH_4 assimilation has been directly or indirectly related to S. Sulfur deficiencies can decrease symbiotic N_2 fixation which can result in plant N deficiencies.

Sulfur deficiency is usually manifested through visual symptoms of light green to yellow leaves which appear first along the veins of young leaves. Legumes are particularly susceptible to S deficiencies and are often diagnosed as having N deficiency because of the similarity in the symptoms. Alfalfa and canola, because of their high S requirements, can be very susceptible to S deficiency. In corn, S deficiencies are generally noted by yellowing of the newer leaves, whereas N deficiencies show up on the older leaves. Sometimes emerging plants will show symptoms of S deficiency, but these symptoms diminish as roots extend further into the subsoil.

6.4.2 Availability Indices of Sulfur

Several methods are available to assess the S status of soils and plants. There are no universal techniques for evaluating S availability in soils, or diagnosing S deficiencies in plants. However, it is known that total S in soils is not a reliable test of S availability since much of the S exists in forms unavailable to plants. For soils, S availability is generally estimated by extracting the soil with water or solutions containing a salt, dilute acid, or phosphate. Depending of the nature of the extracting solution, information can be obtained on the soil content of (1) labile SO_4, (2) labile and adsorbed SO_4 or (3) labile and adsorbed SO_4 and portions of organic S. In addition to extraction methods, incubation techniques and microbial assays have also been used to estimate plant-available S.

Some commonly used techniques for diagnosing S deficiencies in plants include: total S, SO_4-S, ratio of SO_4-S:total S, N:S ratio, and the Diagnosis and Recommendation Integrated System (DRIS). Total S levels in plants are generally an indication of soil S availability since plants acquire most of their S as soil solution SO_4. However, several factors must be considered when interpreting total S data because total S varies with plant parts, age of plant and tissue, and interactions with other elements. Low, sufficient, and high total S levels for a number of forages and agronomic crops are listed in Table 6-8. The amount of plant SO_4-S can be used to estimate when S is deficient or in excess. A low plant SO_4-S content would indicate SO_4 has been incorporated into organic S compounds whereas high SO_4-S contents suggest the plant is accumulating S. The ratio of SO_4-S:total S is relatively independent of plant growth stage and may provide a better estimate of plant S requirements; the critical ratio is about 0.1. Total N:total S ratio has also been used to estimate the status of plant S, but the ratio by itself does not indicate whether N is high or low, or whether S is high or low.

In the DRIS method, several comparisons are made among elemental ratio indices against established norm values. Three advantages to using the DRIS method are (1) analyses are independent of plant age and tissue, (2) nutrients are ranked in order of the most to the least limiting, and (3) nutrient balance is stressed. The DRIS method is a holistic approach to plant diagnostics, and the information derived from this method can be used to evaluate most of the macronutrients and some of the micronutrients. Further discussion on the DRIS systems can be found in Walworth and Sumer (1987), Westerman (1990), and Jones et al. (1991).

Rates of S mineralization are not proportional to the amount of S in organic matter, due to the variety of S-containing organic compounds in soils that can have different decomposition rates, the type of plant and animal residues that can affect mineralization-immobilization rates and release, and the formation of S-containing precipitates (e.g., $CaSO_4$, $[Al_2(SO_4)_3]$) that can influence the amount of plant available S. The C:S ratio of these materials is extremely important since net mineralization occurs when C:S is less than 200:1, net immobilization generally occurs when C:S is greater than 400:1, and steady state results when C:S ratios are >200 and <400:1.

Table 6-8   Total S Requirements in
Various Forages and Agronomic Crops

Crops	S (%)		
	Low	Sufficient	High
Alfalfa	<0.25	0.25–0.50	>0.50
Barley	<0.15	0.15–0.40	>0.40
Clover, white	<0.25	0.25–0.50	>0.40
Corn	<0.25	0.25–0.80	>0.80
Cotton	<0.25	0.25–0.80	>0.80
Grass, brome	<0.17	0.17–0.30	>0.30
Oat	<0.15	0.15–0.40	>0.40
Peanuts	<0.20	0.20–0.35	>0.35
Soybeans	<0.20	0.20–0.40	>0.40
Sugarcane	<0.14	>0.14	—
Wheat, spring	<0.15	0.15–0.40	>0.40

Source: Jones et al., 1991.

6.5 ENVIRONMENTAL MANAGEMENT OF SULFUR

In order to properly manage S, one must first understand its role in the environment. First, we know, with respect to our concern of S in the environment, that acid deposition and acid mine drainage are the two most formidable problems that must be faced. Second, S is an essential plant nutrient element that is receiving more and more attention with the reduction in S-containing materials (i.e., low S fertilizers, use of non-S pesticides, reduced atmospheric S inputs) being added to croplands. Third, S biogeochemistry is only one part of the overall biogeochemistry cycle that comprises many elements. No universal management plan can be designed for S since each situation warrants a close examination of the role S plays in both economic and environment issues. For example, considering our current state of affairs with respect to acidic deposition and its immense publicity, it is not surprising that any problem potentially associated with acid rain will warrant government action. The recent enactment of the Clean Air Act (1991) has specifically designed regulations which call for a reduction in the amount of S that can be released into the atmosphere. Although acid deposition has been shown to be detrimental, especially within areas surrounding the S-emitting source, atmospheric S supplying S to crops has been a clear benefit.

6.5.1 Management of Sulfur in Agriculture

As with N, P, K, Ca, Mg, and the micronutrients, S must also be managed in order to achieve a proper balance between deficiency and excess. A management program must provide sufficient amounts of all plant nutrients in order to obtain economic yields while giving due consideration to potential environmental impacts. Unlike N and P, the agricultural role in cases where S pollution has been identified is at most minimal, if a contribution is identified at all. Therefore, management of S in agriculture is more of a concern in supplying adequate amounts of S for crop production, than is its impact on the environment.

Management of the S supply for agricultural purposes is dependent on the properties of S and the factors that govern its availability. Figure 6-7 describes the various

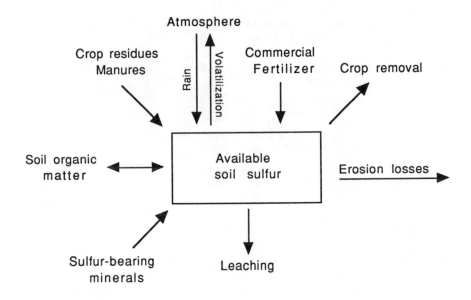

Figure 6-7 Example of S inputs and outputs in an agricultural ecosystem.

S inputs and S outputs that are generally part of an agricultural production setting. Inputs of S such as crop residues and manures, S in rainfall, fertilizers, mineral weathering, and organic matter mineralization contribute to enhancing the soil S supply. Outputs or losses of S include immobilization, volatilization, crop removal, leaching, and erosion. Preventing erosion, returning crop residues, and adding other sources of S (i.e., sewage sludge and composts) will increase the amount of S in soils, in addition to supplying other essential plant nutrients.

As an approximate measure, about 1–3% of the soil organic matter will be mineralized each growing season. The amount of SO_4 released in the mineralization processes will depend on the nature of the organic matter and microbial assimilation. Predicting S needs is therefore difficult since most of the S in soils exists in organic forms that must undergo mineralization for the S to be released. Factors that affect the amount of plant available S include temperature, pH and moisture, all of which directly influence the growth and activity of microorganisms. If crop production uses less S than the amount of S mineralized, a net annual release of nutrients that exceeds plant needs could result in a potential loss of S and decrease available S to future crops. Therefore, it is important to use management practices that balance S needs.

There are many types of S fertilizers, several of which contain other essential nutrients. Examples include elemental S, K_2SO_4, $(NH_4)_2S_2O_3$, $CaSO_4 \cdot 2H_2O$, and $MgSO_4 \cdot 7H_2O$. The choice of one S-containing fertilizer over another should be determined based on cost when comparing fertilizers on an equivalent SO_4-S basis. Factors that should be taken into consideration in a S management program should include: type of fertilizer S, timing of application, method of application, and placement. Both the physical and chemical characteristics of the fertilizer and soil will determine the best method and timing of application. Sewage sludge and manure can also be used as a source of S. The S content (dry weight basis) of sewage sludge from over 200 municipalities in eight states ranged from a low of 0.6% to a high of 1.5% with a median value of 1.1%.

Several criteria can be used in determining the economic benefits derived from S fertilizer use. Some of these criteria include: economic optimum S fertilizer rate, minimum S fertilizer rate, S fertilizer rate to maximize return on investment, minimum cost per unit of yield, and discounting future returns. All but the last criterion are based on present year conditions such as cost of S fertilizer and additional expenses, expected yield, and anticipated return on investment. Discounting for future returns relies on additional benefits due to residual effects. These residuals may include S carryover, mineralization of other nutrients, or prevention of S deficiencies in perennial crops. In a long-term S management program, the latter strategy may be the most beneficial; however, on a short-term basis, the other economic criteria may result in increased profit margins.

6.5.2 Reclamation of Acid Soils and Mine Spoils

When material containing reduced S compounds is exposed to oxygen, high concentrations of sulfuric acid (H_2SO_4) can develop and form acidic soils. Two examples where this might occur are (1) the draining of coastal plains or tidal marsh sediments containing pyritic materials and (2) the exposing of overburden materials to the atmosphere during surface mining activities. Reclamation of these areas requires the neutralization of both the active and the potential acidity that results as oxidation of the reduced S occurs. Neutralization of the potential acidity is an important part of a reclamation program because over time reduced S minerals will be oxidized and contribute to the soil active acidity. In most cases, potential acidity may require greater reclamation efforts than the neutralization of the active acidity. Acid-sulfate soils, commonly referred to as cat clays, develop in drained coastal floodplains located mainly in temperate and tropical regions. After draining these soils, the reduced S compounds oxidize to produce sulfuric acid, which can in turn dissolve mineral matter contained in them. Both acid sulfate soils and mine spoils can contain large amounts of pyrite (FeS_2) which on oxidation can lower soil pH levels to below 2 (see Equation 6-6).

Neutralization of some of the acidity produced during the oxidation of reduced S compounds occurs when silicate minerals dissolve; however, during this process high levels of potentially toxic metals such as Al, Cu, Fe, Mn, Ni, Zn, and others may result. Micronutrient deficiencies involving Mo and B may occur with some vegetation due to the low solubility of these nutrients in acid soils. Reclamation of acid soils and mine spoils requires addition of liming materials to neutralize the acidity and consideration of acid-tolerant crops. Soil tests should be done every 1–2 years, and use of finely ground liming materials is recommended for faster acid neutralization. Reclamation efforts can turn unproductive soils into lands useful for crop production or rangeland use in addition to alleviating a potential source of environmental pollution.

6.5.3 Reclamation of Sodic Soils

Soils of the arid and semiarid regions of the world can accumulate appreciable amounts of salts. However, sodic soils have an exchangeable sodium percentage

(ESP) of 15% or greater and low soluble salt concentrations (<4 dS/m) which results in a greater influence by the Na ions. Under these conditions, clays disperse causing a reduction in soil pore diameter, destruction of soil structure, and decreased soil water permeability. In addition to the major influence of Na on soil physical properties, Na can also be toxic to certain plants when other cation concentrations are low.

Reclamation of sodic soils can be expensive if conditions are to the point where water infiltration and permeability are very slow. The exchangeable Na must be replaced by other cations, preferably Ca, which is divalent and readily replaces Na. Gypsum ($CaSO_4$) is very soluble and once dissolved can supply the necessary Ca needed to displace Na. An application of several tons of gypsum may be required per hectare in order to supply enough Ca. The amount of gypsum required (GR) for reclamation can be approximated from the ESP and CEC of the sodic soil. For example, if the sodic affected area has an ESP of 25% and a CEC of 20 cmol/kg, and the goal is to lower the ESP to 5%, the GR would be as follows:

$$GR = (Na_x)*4.5 \text{ metric tons/ha to a depth of 30 cm} \qquad (6\text{-}8)$$

where Na_x represents the cmol Na/kg to be replaced (i.e., 20% of 20 cmol/kg). Thus, a total of 18 metric ton/ha will be required to replace the 4 cmol Na per kilogram. For best results the gypsum should be incorporated into the surface or subsurface. Adequate water must also be applied to leach out the Na that was displaced by the Ca ions.

REFERENCES

Brady, N. C., *The Nature and Properties of Soils,* Macmillan, New York, 1974.

Freney, J. R., Forms and reactions of organic sulfur compounds in soils, in *Sulfur in Agriculture,* Agronomy Monograph Number 27, Tabatabai, M. A., Ed., American Society of Agronomy, Madison, WI, 1986, 207.

Goodrich, R. D. and Garrett, J. E., Sulfur in livestock nutrition, in *Sulfur in Agriculture,* Agronomy Monograph Number 27, Tabatabai, M. A., Ed., American Society of Agronomy, Madison, WI, 1986, 617.

Jones, J. B., Jr., Wolf, B., and Mills, H. A., *Plant Analysis Handbook: A Practical Sampling, Preparation, Analysis, and Interpretation Guide,* Micro-Macro Publishing, Athens, GA, 1991.

Krupa, S. V. and Tabatabai, M. A., Measurement of sulfur in the atmosphere and in natural waters, in *Sulfur in Agriculture,* Agronomy Monograph Number 27, Tabatabai, M. A., Ed., American Society of Agronomy, Madison, WI, 1986, 251.

Landers, D. H., David, M. B., and Mitchell, M. J., Analysis of organic and inorganic sulfur constituents in sediments, soils and water, *Intern. J. Environ. Anal. Chem.,* 14, 245, 1983.

Mitchell, M. J., David, M. B., and Harrison, R. B., Sulphur dynamics of forest ecosystems, in *Sulfur Cycling in Terrestrial Systems and Wetlands,* Howarth, R. W., and Stewart, J. W. B., Eds., John Wiley & Sons, New York, 1992.

Stevenson, F. J., The sulfur cycle, *Cycles of Soil: Carbon, Nitrogen, Phosphorus, Sulfur, Micronutrients,* John Wiley & Sons, New York, 1986, chap. 8, 285.

Thompson, J. F., Smith, I. K., and Madison, J. T., Sulfur metabolism in plants, in *Sulfur in Agriculture,* Agronomy Monograph Number 27, Tabatabai, M. A., Ed., American Society of Agronomy, Madison, WI, 1986, 57.

Tisdale, S. L., Nelson, W. L., and Beaton, J. D., *Soil Fertility and Fertilizers,* Macmillan, New York, 1985.

Trudinger, P. A., Chemistry of the sulfur cycle, in *Sulfur in Agriculture,* Agronomy Monograph Number 27, Tabatabai, M. A., Ed., American Society of Agronomy, Madison, WI, 1986, 1.

Walworth, J. L. and Sumner, M. E., The diagnosis and recommendation integrated system, *Adv. Agron.,* 194, 1987.

Westerman, R. L., Ed., *Soil Testing and Plant Analysis,* SSSA Book series number 3, Soil Science Society of America, Madison, WI, 1990.

Williams, C. H., The chemical nature of sulphur in some New South Wales soils, in *Handbook of Sulphur in Australian Agriculture,* McLachlan, K. D., Ed., CSIRO, Melbourne, Australia, 1974, 66.

SUPPLEMENTARY READING

Brimblecombe, P. and Lein, A. Y., Eds., *Evolution of the Global Biogeochemical Sulphur Cycle,* John Wiley & Sons, New York, 1989.

Paul, E. A. and Clark, F. E., *Soil Microbiology and Biochemistry,* Academic Press, San Diego, 1989.

Schlesinger, W. H., *Biogeochemistry: An Analysis of Global Change,* Academic Press, New York, 1991.

Tabatabai, M. A., Ed., *Sulfur in Agriculture,* Agronomy Monograph Series No. 27, American Society of Agronomy, Madison, WI, 1986.

7

TRACE ELEMENTS

7.1 INTRODUCTION

Trace elements are elements that are normally present at relatively low concentrations in soils or plants. They may or may not be essential for the growth and development of plants, animals, or humans. Micronutrients and heavy metals are two terms that are sometimes used to describe categories of trace elements. The use of micronutrients is avoided here because the term implies that the elements in question are essential for the growth and development of some organism and many trace elements are not. The use of heavy metals is avoided because the term generally refers only to metallic elements with an atomic weight greater than that of iron (55.8 g/mol) or to elements with a density greater than 5.0 g/cm^3, which also excludes many other trace elements. Other terms that have been used to describe trace elements include trace metals, microelements, minor elements, and trace inorganics. Note that some elements are typically present at high concentrations in soils or the earth's crust, but are considered trace elements because they occur at low concentrations in plants. Titanium, Fe, and Al are three examples.

With such a broad definition, trace elements obviously include a large number of elements with widely varying chemical characteristics. Figure 7-1 is the periodic table of elements with elements that are not considered trace elements shaded. Shaded elements are present in relatively high concentrations in plants and soils and therefore do not fit the definition of trace elements, have existed only as radioactive isotopes that have since decayed to daughter products, do not occur naturally in the environment, or are present only as inert gases. A simple means of reducing the number of elements to be considered would be to limit the discussion to those elements that are of concern because they are either essential or readily toxic to humans, animals, or plants. By this method we might consider As, B, Be, Cd, Co, Cr, Cu, F, Fe, Hg, I, Mn, Mo, Ni, Pb, Se, Sn, V, and Zn. A description of the behavior of all these elements in soils is not possible here although some will be used as examples for general principles.

Trace elements in soils also can be divided into three broad categories based on their expected chemical form in soil solutions. *Cationic metals* are metallic elements for which the predominant form in the soil solution would be a cation. Examples are Ag^+, Cd^{2+}, Co^{2+}, Cr^{3+}, Cu^{2+}, Hg^{2+}, Ni^{2+}, Pb^{2+}, and Zn^{2+}. *Oxyanions* are elements that are combined with oxygen in molecules with an overall negative charge. Examples

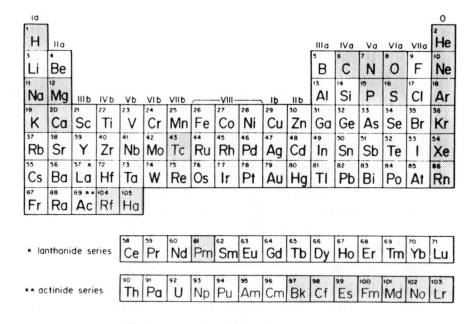

Figure 7-1 Periodic table of the elements. Shaded elements are not considered as trace elements.

are AsO_4^{3-}, $B(OH)_4^-$, CrO_4^{2-}, MoO_4^{2-}, $HSeO_3^-$, and SeO_4^{2-}. The *halogens* are members of group VIIA in the periodic table and are present as anions in the soil solution. They are F^-, Cl^-, Br^-, and I^-. The categories are not mutually exclusive, however, because some elements can occur in more than one category.

A question that is commonly asked is what typical trace element concentrations are found in soils. Usually this is in reference to a potential contamination problem, and an individual wants to know whether the results from a soil analysis reflect a contaminated situation. Unfortunately, there is not a straightforward answer to the question because the natural variability in soil trace element concentrations is quite high. Table 7-1 represents some normal soil concentrations for selected trace elements and some geochemically anomalous concentrations. A soil with a total Pb

Table 7-1 Selected Trace Element Concentrations in Soils at Normal and Geochemically Anomalous Levels

Element	"Normal" range (mg/kg)	Metal-rich range (mg/kg)
As	<5–40	Up to 2,500
Cd	<1–2	Up to 30
Cu	2–60	Up to 2,000
Mo	<1–5	10–100
Ni	2–100	Up to 8,000
Pb	10–150	10,000 or more
Se	<1–2	Up to 500
Zn	25–200	10,000 or more

Source: Bowie and Thornton, 1985.

Figure 7-2 Abandoned Pb and Zn mining site in northeastern Oklahoma.

concentration of 600 mg/kg, for example, could represent a situation in which a soil with an original Pb concentration within the normal range was contaminated or a geochemically anomalous situation. Additional information is required to make the correct determination.

Uses for trace elements are quite numerous. Considerable interest lies in the area of dietary supplements for humans and livestock since some trace elements are essential elements. Pesticides used to commonly contain Pb and As; and Pb is used in gasoline, paints, and plumbing fixtures, but its use has been and continues to diminish rapidly. Zinc is used extensively in manufacturing rubber products and in galvanizing metals. Chromium is used in chrome plating of automobile components and in manufacturing steel.

With such varied uses for trace elements, there are equally varied means for soil contamination. Historically, the mining and smelting of the trace elements themselves has created trace element soil contamination problems of the greatest magnitude. Figure 7-2 is a scene in northeastern Oklahoma showing some of the environmental problems associated with an area once the site of Pb and Zn mining. The material in the background is chat, a waste rock that can have Zn and Pb concentrations as high as 2%. Fine particles selectively erode away from the chat piles contaminating nearby soils and becoming sediments in surface waters, greatly enlarging the area impacted by the original mining activity. In addition, these chat piles are a source of dust that can expose people to Pb across a wide area. Mining activities such as this produce primary contaminants consisting of waste rock, tailings, and slag. Secondary contamination occurs in groundwater beneath open pits and ponds, sediments in river channels and reservoirs, floodplain soils impacted by contaminated sediment, and soils affected by smelter emissions. River sediments reworked from floodplains and groundwater from contaminated reservoir sediments are tertiary contaminants.

The mining and use of other materials, such as coal, often lead to trace element soil contamination as well. Municipal wastewater treatment plants that treat domestic and industrial inputs and that use land application as a sludge disposal option may enrich soils with trace elements. Motor vehicles emit Cd from diesel fuel; Zn and Cd from tire attrition; Ni, Cr, V, W, and Mo from attrition of steel; and Pb from gasoline (at a much reduced rate in the United States). Fossil fuel residues are also a source of trace elements as are some agricultural sprays and soil amendments. Pigs fed Cu as a dietary supplement will produce Cu-enriched manure which can then be applied to agricultural soils. Urban areas will generally have high soil trace element concentrations due to their proximity to more trace element sources. Figure 7-3 shows a general decline in Cu, Ni, and Pb concentrations in forest floor and soil samples with distance from the center of New York City, illustrating the decline in trace element concentrations with distance from a generalized source.

7.2 ADVERSE EFFECTS

It is difficult to summarize all adverse effects from trace elements because of the large number of elements and because each element can cause a number of different problems. In general, the concerns are *human health, animal health, phytotoxicities,* and *impacts on aquatic ecosystems.* Of particular importance is the fact that, with few exceptions, trace element contamination of soils is, for all practical purposes, permanent. One cannot easily remove most of the trace elements from a soil once they have been added. A few examples are provided below to provide the reader with an appreciation for the types of problems that are of concern.

7.2.1 Human Health

Humans are exposed to trace elements in soil through food chain transfer and by direct ingestion of soil particles (Figure 3-1). Documented cases of acute trace metal toxicities in humans due to elevated soil trace element concentrations are rare or nonexistent. Documented cases of adverse health effects due to chronic food chain exposure are more numerous. An often cited example deals with Japanese farmers suffering from Cd toxicity after long-term consumption of Cd-enriched rice. Two symptoms of Cd toxicity are renal dysfunction and *itai-itai* disease. Itai is Japanese for "ouch," and the disease is characterized by Cd-induced bone loss which produces localized, severe pain in the victims. The rice had been grown in paddies polluted by Pb and Zn mining and smelting activities. There is a geochemical association between Zn and Cd, and therefore Zn-contaminated soils are often Cd contaminated as well. Table 7-2 indicates that rice grown in areas having >10% morbidity had Cd concentrations as much as 14 times higher than rice grown in nonpolluted areas.

Efforts to reduce or eliminate Pb and gasoline in paints were initiated because of the indirect inhibition of heme production in humans by Pb. Some health officials believe that chronic Pb exposure can cause mental impairment, particularly in

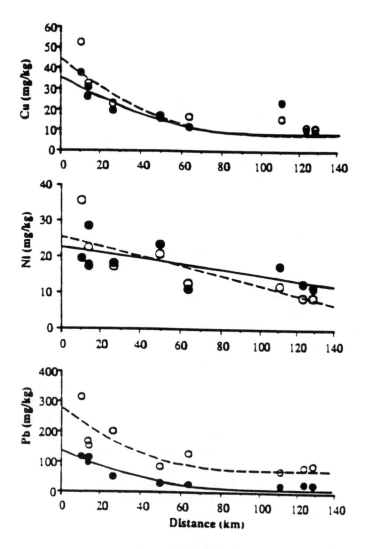

Figure 7-3   Forest floor and soil heavy metal concentrations as a function of distance from Central Park, Manhattan, New York City. Open circles represent forest floor and closed circles represent soil values. (Reprinted from Pouyat and McDonnell, 1991. With permission.)

children, and that exposure to soil Pb should be strictly controlled. Dietary studies of people living downwind from old Zn and Pb smelter sites have shown that persons could — by consuming some of their home-grown vegetables, meat, and milk — ingest at least 50% more Pb and Cd than if comparable food items had been purchased in a control area. Recent studies on the effects of Pb on human health that indicate significant health effects can occur at lower Pb doses than previously believed, have greatly increased interest in Pb-contaminated soils. Older urban areas will typically have elevated soil Pb concentrations due primarily to the use of leaded house paints. Children are exposed by directly consuming paint chips or soil.

Table 7-2   Cadmium Concentrations[a] in Rice
by Prevalence of Itai-Itai Disease

	Area	
Rice type	High endemic[b]	Nonendemic
Nonglutinous		
Polished	0.52	0.048
Unpolished	0.54	0.079
Glutinous		
Polished	1.03	0.071
Unpolished	1.12	0.150

Source: Tsuchiya, 1978.
[a] mg/kg.
[b] >10% Morbidity.

Of increasing interest recently is the possibility of subclinical effects due to exposure to slightly elevated amounts of trace elements. Subclinical effects mean symptoms or diseases not diagnosed by medical examination, for example, attention deficit disorder and hyperactivity in children. Epidemiological studies are required to detect statistically significant trends. The number of people affected by trace element poisoning worldwide is quite astounding, as shown in Table 7-3. Also note that soils are the primary recipient of global emissions of trace elements.

7.2.2 Animal Health

Toxicities of Mo in livestock *(molybdenosis)* and Se in livestock and wildlife *(selenosis)* are two relatively common trace element animal health problems. Ruminant animals are the most susceptible to molybdenosis, a Mo-induced Cu deficiency. Some soils have naturally high Mo concentrations and produce forages that can induce molybdenosis in grazing animals. Soils can be naturally enriched in Mo or become enriched because of additions of poor quality irrigation water or Mo-rich sewage sludge. Selenium problems are much more widespread. Outcroppings of seleniferous sedimentary rocks occur in certain regions throughout the United States, and particularly in semiarid to arid regions where the associated soils may produce plants with Se concentrations that contain Se levels suspected to be harmful to livestock. Coal mining activities may also bring seleniferous materials to the surface. The Kesterson Wildlife Refuge area of central California represents a situation where agricultural activities have led to some serious outbreaks of Se poisoning to wildfowl. Irrigated cropland produced drainage water with a high Se concentration; and as this water flowed into the Kesterson area, it was further concentrated by evaporation and produced widespread Se toxicity problems in aquatic life and waterfowl.

Table 7-3   Estimated Magnitude of the Extent of Trace Element Poisonings

	Global emissions (1000 mt/yr)			People	
Element	Air	Water	Soil	affected	Comments
Pb	332	138	796	>1 billion	Blood Pb > 20 µg/dL
Cd	7.6	9.4	22	500,000	Producing renal dysfunction
Hg	3.6	4.6	8.3	80,000	Certified Hg poisonings
As	18.8	41	82	>100,000	Skin disorder and H_2O As >2 µg/L

Source: Nriagu, 1988.

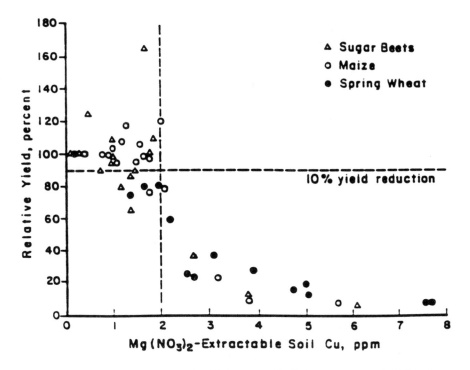

Figure 7-4   Relationships between Mg (NO₃)₂-extractable Cu in soil and crop yield. (Reprinted from Lexmond and deHaan, 1977. With permission.)

7.2.3 Phytotoxicities

Phytotoxicity problems are of concern for two primary reasons. First, the potential productivity of cropland can be reduced. An example is illustrated in Figure 7-4. Here the relative yield of three crops declines as the $Mg(NO_3)_2$-extractable soil Cu concentrations increase. When extractable Cu concentrations exceed that shown by the vertical dashed line, relative yield is predicted to decline by >10%. In this situation, Cu was being studied because of potential phytotoxicities from Cu additions via swine manure. Manure amended soils had not, in general, reached the critical value of 2 mg Cu per kilogram. Second, in situations where vegetation is sparse because of trace element phytotoxicities, wind and water erosion can occur at an accelerated rate. These conditions often exist around sites where metal mining or smelting once took place (Figure 7-2). The lack of vegetative cover allows further dispersal of the trace elements into the surrounding environment, further aggravating the problem.

7.2.4 Aquatic Environments

The effects of elevated concentrations of trace elements in aquatic environments are extremely difficult to assess, due in part to the mobility of some of the species and to the difficulty in separating out the effects of contaminated water from contaminated sediment. Soil erosion is the primary mechanism by which trace elements are transferred from soils to the aquatic environment. The enrichment ratios for most trace elements are ≥1. Aquatic species can be divided into the groups of plants

Table 7-4 A Summary of Species Most Commonly Affected by Toxicities of Selected Trace Elements

| Element | Species adversely affected | | | | |
	Humans	Animals	Aquatic organisms	Birds	Plants
Cd	*	*	*	*	*
As, Pb, Hg, Cr, Se	*	*	*	*	
Cu, Ni, Zn			*		*
Mo, F, Co		*			
B					*

Source: Page, 1992.

(phytoplankton and benthic), invertebrates, and fish. The effects of trace elements are generally manifested by a reduction in the diversity, productivity, and density of aquatic organisms.

Any trace element can have an adverse effect on any organism if the dose is high enough. Exposure of some organisms to high doses of certain trace elements is uncommon and, therefore, is not considered an environmental problem. Table 7-4 indicates the species groups that are most commonly at risk due to exposure to elevated doses of 13 important trace elements. Note, for example, that Pb is of concern for humans, animals, aquatic organisms, and birds but rarely induces phytotoxicities (except in unique, highly contaminated situations) as compared to B for which phytotoxicities are of primary concern.

7.3 BIOAVAILABILITY OF TRACE ELEMENTS IN SOILS

The scope of this chapter has been on excessive concentrations of trace elements in soils. For the problems of phytotoxicities and food chain transfer, the topic of bioavailability becomes important since it may be desirable to estimate or even reduce the bioavailability of trace elements.

As was described in Chapter 2, bioavailability is the possibility that a substance in the environment (trace elements in soils in this instance) will cause an effect, either positive or negative. Total trace element concentrations in a soil are, in fact, poor predictors of trace element bioavailability. The bioavailable fraction is some subfraction of the total trace element fraction that best predicts the response. Here, this response generally refers to plant uptake of the trace element. It must be recognized, however, that trace element additions to soils, which will increase total trace element concentrations, will likely increase trace element bioavailability at least to some extent. In this situation, a relationship may exist between total and bioavailable trace element concentrations.

Copper can be used as an example to illustrate some of the processes involved in bioavailability. Beginning with the soil solution, one could easily measure the total Cu concentration in the solution (not to be confused with the total Cu concentration in a soil sample) with atomic absorption spectrophotometry. This concentration is denoted Cu_T, and in fact it represents a summation of many different soluble Cu species, as indicated below:

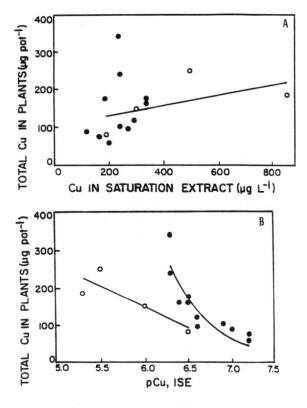

Figure 7-5   Relationship between total Cu in snapbean plants and Cu_T (a) or pCu (b) in the soil solution. Open circles represent Cu additions to the soil from $CuSO_4$ and filled circles represent Cu additions from sewage sludge. (From Minnich et al., 1987. With permission.)

$$Cu_T = Cu^{2+} + CuOH^+ + CuCO_3^0 + CuCl^+ + ... \qquad (7-1)$$

As Figure 7-5 implies, it is the Cu^{2+} concentration (more precisely, the Cu^{2+} activity) that determines the Cu bioavailability. Figure 7-5A shows the relationship between Cu_T and the Cu concentration in snapbean plants. The filled circles represent Cu additions to the soil from sewage sludge, and the open circles represent Cu additions from $CuSO_4$. Figure 7-5(b) shows the relationship between pCu (negative logarithm of the Cu^{2+} activity; as pCu increases, the Cu^{2+} activity decreases) — as measured with an ion selective electrode — and the Cu concentration in snapbean plants. Qualitatively one can see that pCu is a better predictor of Cu concentrations in the plants than Cu_T. Generally, each trace element has a soluble species (Cu^{2+} in this example) that is preferentially taken up by plants. Thus one key to bioavailability is related to the factors that control the activity of soluble trace element species in the soil solution that is preferentially taken up by plants. Intuitively, this should make sense because roots obtain Cu from the soil solution.

Proceeding with our Cu example, the free Cu^{2+} ion can undergo adsorption reactions, precipitate as some Cu-containing solid phase, interact with organic matter, form other soluble species, or change oxidation state if the oxidation/reduction (redox) conditions of the soil change.

The adsorption reactions for cations generally involve interactions with the cation exchange sites where, in this example, the electrostatic attraction between the Cu^{2+} ion and the negatively charged exchange sites results in adsorption of the Cu^{2+} onto the soil solids. Since soils generally have a net negative charge, this interaction is obviously of greater importance for the cationic metals. Anion adsorption can also involve electrostatic forces, but the adsorption process is generally characterized by the formation of a chemical bond between the trace element and the soil solids. Molybdate adsorption onto Fe or Al oxide minerals is an example. Adsorption reactions for the oxyanion trace elements are analogous to those described in Section 5.3.1 for $H_2PO_4^-$ and HPO_4^{2-} and Section 6.3.1 for SO_4^{2-}.

Precipitation of trace element solid phases occurs when the solubility of a solid phase is exceeded in the soil solution, as illustrated in the reaction between two trace elements shown below:

$$PbMoO_4 \rightleftharpoons Pb^{2+} + MoO_4^{2-} \tag{7-2}$$

If the soil system were in equilibrium with $PbMoO_4$ (wulfenite), then the solid phase might control the activities of Pb^{2+} and/or MoO_4^{2-} in the soil solution.

Soil organic matter contains several types of functional groups that can complex trace elements or act as exchange sites. An example reaction in which carboxyl (COOH) and hydroxyl (OH) groups act together to complex Cu^{2+} is shown below:

Several trace elements can undergo changes in oxidation state under the range of redox conditions found in soils. Arsenic, Cu, Cr, Fe, Mn, and Se are typical examples. Selenium, for example, can exist as selenide (Se^{2-}), elemental Se (Se^0), selenite (SeO_3^{2-} with Se^{4+}), and selenate (SeO_4^{2-} with Se^{6+}) in soils and sediments. Selenate has a much higher bioavailability than the more reduced forms of Se. The redox conditions of the soil can also indirectly affect trace element bioavailability. For example, Cd bioavailability is lower under flooded (anaerobic) conditions. This may be due to the formation of insoluble CdS in the reducing environment.

The solubility and as a consequence the bioavailability of trace elements can be greatly influenced by soil pH. Generally, the bioavailability of the cationic metals will increase as soil pH decreases. Soil pH effects on the bioavailability of oxyanions are more variable. Arsenic, Mo, Se, and some forms of Cr often become more available as pH increases.

There are a number of reasons why the bioavailability of trace elements changes with changes in soil pH. Protons or hydroxyl ions can compete for adsorption or

Table 7-5 Concentrations[a] of Selected Elements in Alfalfa Tissue as Influenced by Soil pH

pH	Cd	B	Cu	Ni	Mo
6.0	0.8	64.1	17.7	1.9	193
7.0	0.6	40.9	16.8	0.8	342
7.7	0.4	31.8	16.0	0.8	370

Source: Pierzynski, 1985.
[a] mg/kg.

complexation sites, and therefore a change in the activity of H^+ or OH^- can alter the partitioning of trace elements between the soil solution and the soil solids. A change in the soil solution pH can change the suite of trace element containing soluble species. For example, at pH <6 Pb^{2+} predominates while at pH 6–11 the $PbOH^+$ form predominates. Trace element solid phases can also have pH-dependent solubilities. In situations where these solid phases are controlling the solubility of a trace element, the bioavailability of the element will change with changes in pH. A simple example is given below where the mineral tenorite (CuO) is dissolved by the addition of hydrogen ions. The equilibrium expression is also given along with an algebraic expression describing the changes in the logarithm of the Cu^{2+} activity with changes in pH. In deriving the last expression we have assumed that the activities of H_2O and the solid mineral CuO are equal to one, have taken the logarithm of the equilibrium expression, and have converted log (H^+) to pH.

$$CuO + 2H^+ \rightleftharpoons Cu^{2+} + H_2O \tag{7-4}$$

$$\frac{\left(Cu^{2+}\right)\left(H_2O\right)}{\left(CuO\right)\left(H^+\right)^2} = 10^{7.66} \tag{7-5}$$

$$\log \left(Cu^{2+}\right) = 7.66 - 2pH \tag{7-6}$$

Thus as the pH decreases, the Cu^{2+} activity will increase and vice versa.

Table 7-5 illustrates the influence of soil pH on B, Cd, Cu, Mo, and Ni concentrations in alfalfa tissue. The bioavailability of B, Cd, Cu, and Ni decreases with increasing pH while the opposite is true for Mo.

In practice it is extremely difficult or impossible to determine ion activities in a sample that is representative of the soil solution. There are ion selective electrodes available for a few ions that can directly measure ion activities in aqueous solutions. Even in this situation, the activities found in soil solutions or extracts are often below the detection limit of the electrode unless one is dealing with a highly contaminated soil. Other less direct and more cumbersome methods exist for estimating ion activities, but they are not applicable to routine soil analysis. As a substitute, various soil extractants have been proposed that empirically estimate the bioavailability of trace elements. In the simplest cases, these extractants are weak salt solutions that will have trace element concentrations higher than those found in the soil solution (circumventing detection limit problems). The data shown in Figure 7-4 are examples. More complex fractionation procedures that attempt to separate total soil trace element concentrations into exchangeable, water soluble, carbonate or oxide

Table 7-6   Concentrations[a] of Cd and Zn in Leaf Tissue of Various Crops as Influenced by Soil Series

Plant	Cd			Zn		
	Gazos[b]	Greenfield	Domino	Gazos[c]	Greenfield	Domino
Tomato	5.8	16.5	12.9	20.0	121.0	99.5
Pepper	26.2	11.4	10.2	108.0	195.8	205.0
Leaf lettuce	21.9	63.7	15.8	58.0	354.0	137.0
Head lettuce	26.3	41.6	17.9	57.3	281.0	143.8
Radish	19.6	17.9	14.0	48.8	332.8	246.8
Potato	15.6	8.7	4.6	26.5	50.5	27.5
Corn	4.1	19.3	14.1	47.0	239.5	154.8
Wheat	1.8	9.0	3.3	42.5	210.0	130.5
Swiss chard	8.0	37.9	19.5	33.5	884.0	349.8
Broccoli	7.4	8.8	6.4	19.0	288.0	167.8
Carrot	10.7	14.5	10.5	47.8	155.5	109.5
Beet	11.5	51.8	23.7	48.3	1023.3	446.5

Source: Kim et al., 1988.
[a] mg/kg.
[b] Total Cd concentrations were 5.6, 5.1, and 5.3 mg/kg for the Gazos, Greenfield, and Domino soils, respectively.
[c] Total Zn concentrations were 147, 250, and 309 mg/kg for the Gazos, Greenfield, and Domino soils, respectively.

bound, organic, precipitated, or residual forms have also been developed. For these extraction schemes to be useful, a relationship similar to that shown in Figure 7-5(b) must exist between extractable trace element concentrations and plant trace element concentrations.

Plants themselves may be used as indicators of trace element bioavailability. Plant responses to trace elements in soils are as *accumulators, indicators,* or *excluders.* Accumulators are plants that actively take up an element, out of necessity or with no apparent need, and can have extremely high concentrations of the element without suffering any ill effects. Legumes can accumulate Mo even though their metabolic requirements for Mo are quite low. Some species in the genera *Astragalus* are Se accumulators, and their presence can be used to identify Se-rich soils. Indicators are species that take up an element in proportion to the amount present in the soil (until phytotoxicity occurs) but do not necessarily require it for their growth. Excluders actively exclude elements such that the concentration of the trace element in the plant will remain relatively low and constant at varying soil trace element levels.

Table 7-6 illustrates the variation in Cd and Zn concentrations that can be found in various plant species grown in the same soil and the variation in Cd concentrations for a given plant species grown on three soils with comparable Cd concentrations. There is as much as a 15-fold difference in Cd concentrations across plant species within soils and as much as a 5-fold difference in Cd concentrations for a given plant species across soils. It is noteworthy that plants grown in the Gazos soil, which has the highest pH and cation exchange capacity, do not consistently have lower Cd concentrations compared to plants grown in the remaining soils.

A basic understanding of the processes involved in trace element bioavailability is helpful in understanding the reasoning behind the management options for trace element contaminated soils. These processes are summarized in Figure 7-6 which depicts a generalized biogeochemical cycle for trace elements in agroecosystems. The total amount of a trace element in the soil solution at a given time is small by

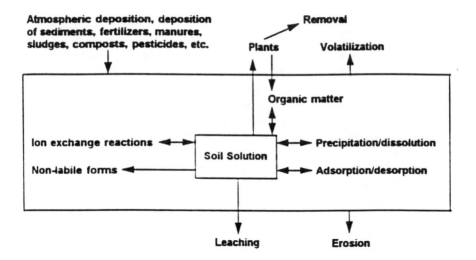

Figure 7-6 Generalized trace element cycle in soils. The large rectangle defines the boundaries of the solum.

comparison to what plants remove over time, and there is a dynamic interaction between the soil solution and the soil solids. Processes interacting with the soil solution that have a bidirectional arrow will be in rapid equilibrium with the soil solution. Such pools of trace elements are sometimes called *labile* and likely represent the most bioavailable forms of the trace elements. Rapid implies a time frame, and various reference points have been used. The most useful would be that of plant uptake. As plants remove trace elements from the soil solution, the labile pools can replenish the soil solution in a sufficiently short period of time such that the plant has an essentially constant source of the trace element. *Nonlabile* forms would contribute very little to plant uptake. Nonlabile forms include trace elements physically isolated from the soil solution (occluded) and sparingly soluble trace element solid phases. Note that there are only four ways for trace elements to be removed from the soil system — crop removal, erosion, leaching, and volatilization. Crop removal is generally insignificant by comparison to the quantities of trace elements accessible by the crop root system. Erosion and leaching losses are undesirable since they represent additional dispersal of a potential pollutant. Volatilization is an important pathway for only a few trace elements.

7.4 REMEDIATION OPTIONS FOR TRACE ELEMENT-CONTAMINATED SOILS

As with many problems, prevention is usually the best recourse and trace element contamination of soils is no exception. Recall that it was stated that trace element contamination of soils should be considered permanent in most cases. In light of this, trace element additions are sometimes regulated. A case in point would be the recent sewage sludge disposal regulations which have limits on the total trace element additions that can be made to soils via land application of sludge. Those limits are given in Table 9-4. The limits were set after the appropriate risk assessment procedures (to be described in Chapter 11) were followed and a compromise was made between protecting soil quality and utilizing the soil as a disposal medium for the plant nutrients and organic matter contained in the sludge.

There are two general strategies that can be taken with trace element-contaminated soils. The first, which we shall collectively term *treatment technologies*, refers to soil that has been physically moved and processed in some fashion in an attempt to reduce the concentrations of trace elements or to reduce the *toxicity characteristic leaching procedure* (TCLP) extractable trace element concentrations to an acceptable level. The TCLP is a protocol used by the U.S. Environmental Protection Agency (EPA) which dictates that materials are leached under standard conditions. If the concentrations of various substances exceed some critical level in the leachate, the material is classified as hazardous. The second, which we shall term *on-site management*, refers to soil that is managed or treated in place. There are two subcategories within the on-site management option. *Isolation* is one of a number of processes by which a volume of soil is solidified and any further interaction with the environment is prevented. The second subcategory consists of methods for *reducing the bioavailability* of trace elements.

7.4.1 Treatment Technologies

The treatment technologies include a wide variety of processes ranging from high temperature treatments that produce a vitrified, granular, and nonleachable material to the addition of solidifying agents that produce a cement-like material to washing processes. A detailed description of these technologies is beyond the scope of this text. The processes are effective, however, either in reducing the trace element concentrations in the soil or in reducing the TCLP extractable trace element concentrations to an acceptable level.

Several drawbacks to the treatment technologies bear mentioning. Conservation of mass applies and any trace elements removed from the soil will still have to be disposed of. This may include additional treatment of large volumes of wastewater. Thus, the trace element problem may have changed form but it still remains. The most significant drawback is the fact that trace element contaminated soil problems frequently involve large quantities of soil. A case in point would be in Cherokee County, Kansas, which has numerous abandoned Pb and Zn mining sites. The soil survey for the county reports 1,316 ha of mine dump sites. All of these areas have very high Pb and Zn concentrations in the soil or mine spoil material located on site and would benefit from remediation. If only the top 15 cm of these areas were treated, it would require that approximately 2.4×10^6 Mg of material be treated. This is a very conservative estimate because treating the top 15 cm would not be sufficient, and there are also many other areas needing remediation that are not readily apparent in the soil survey. Even with an effective treatment technology available, it would be cost prohibitive to treat the quantities of material necessary to address the problem. Thus, the treatment technologies are confined to situations where relatively small volumes of soil need to be handled.

7.4.2 On-site Management

The isolation subcategory of the on-site management option includes in-situ applications of processes described under the treatment technologies option. In-situ vitrification utilizes electric current passed through a soil volume which produces a vitrified monolith. In-situ solidification involves mixing the soil with the solidifying agents without removing the soil from its original resting place. The monolith forms in place, similar to the curing of concrete. Both of these processes essentially isolate the contaminants from the surrounding environment by encapsulating them in a nonporous matrix.

Of greatest interest in the context of this book are the methods for reducing trace element bioavailability. These are more the concern of the soil scientist whereas the previously mentioned options are engineering approaches. The methods that will be discussed include *altering soil pH, increasing sorption capacity, precipitation of trace elements* as some insoluble solid phase, *attenuation, metal-P interactions,* and *enhancing volatilization losses.* The influences of soil pH, cation exchange capacity, and adsorption reactions on trace element bioavailability are well studied and reported in soil literature, although generally not in the context of a remediation technique. The remaining methods are often mentioned in engineering writings dealing with soil remediation techniques, but are not always presented with data verifying their effectiveness.

Of all the methods for reducing trace element bioavailability, increasing the soil pH with lime (generally to a pH of 6.5 or greater) is probably the most common method employed. This is because of the general prevalence of problems with cationic metals versus oxyanions and of the fact that soil pH management is a routine part of a soil fertility program. Obviously, if one is dealing with a situation involving more than one trace element, as shown in Table 7-5, changing the soil pH may improve matters for some elements and make matters worse for others.

Studies have shown that the plant availability of some trace elements is influenced by the soil cation exchange capacity, with availability decreasing as the exchange capacity increases. In theory, increasing the cation exchange capacity of a soil is a potential method for reducing trace element bioavailability. An increase in soil cation exchange capacity can be brought about by the addition of clays having a high exchange capacity themselves, or possibly by the addition of organic materials (e.g., manures, sludges, peats). Organic materials can also complex trace elements. With either type of material, soil sorption capacity is increased, which reduces the trace element bioavailability. The adsorption capacity of a soil for oxyanions can also be increased, with a subsequent reduction in their bioavailability, by the addition of large amounts of Fe or Al salts. The salts will dissolve and produce Fe and Al oxides and hydroxides, which are very effective sinks for certain oxyanions. There is some uncertainty as to whether the changes in bioavailability brought about by adding materials such as clays, peats, or Fe or Al salts are due to changes in sorption capacities or to changes in other soil chemical properties such as pH.

The addition of hydroxide, carbonate, phosphate, sulfate, or sulfide-containing salts can cause precipitation of the corresponding trace element-containing solid phase. If the solid phase then controls the activity of the trace element in the soil solution and the activity is lower than the initial level, the bioavailability of the trace element will be reduced.

Attenuation simply refers to the dilution of a contaminated soil with a noncontaminated material. Often this material will be uncontaminated soil but it can also be materials such as sewage sludges, paper mill wastes, animal manures, or coal fly ash. Attenuation can also be obtained by deep tillage, assuming subsurface soil horizons have a lower level of contamination compared to the surface soil material. This is often the case when aerial deposition of trace elements was a factor in the contamination process.

The metal-P interactions have largely been studied with respect to P-induced trace element deficiencies in crops. This P-induced reduction in trace element bioavailability may also be useful in reducing phytotoxic levels of trace elements or in reducing food chain transfer of trace elements. This interaction is prevalent with Zn, Cu, and Fe; however, the exact mechanism of the metal-P interaction is not known. Postulated mechanisms include translocation relationships, metabolic interferences, depressed trace element concentrations in plants due to growth response to P with subsequent dilution of the trace elements, and formation of insoluble trace element solid phases. Phosphorus oxyanions can also interact with other oxyanions. The bioavailability of some oxyanions, particularly As, increases as P levels increase. This interaction is probably due to the replacement of As by P on adsorption sites.

It has long been recognized that some trace elements exist in volatile forms. This is particularly true for As, Hg, and Se. The volatile forms are generally methylated. Examples are trimethylarsine $[(CH_3)_3As]$, mono- or dimethyl mercury $[CH_3Hg^+,$ $(CH_3)_2Hg]$, or dimethylselenide $[(CH_3)_2Se]$. The production of methylated forms of these trace elements is microbially mediated utilizing a variety of substrates. Recently, the idea was put forth to enhance the production of volatile forms of trace elements as a means of remediating contaminated soils. This is one of the few biological methods available for trace elements. Field studies have shown close to 70% reduction in Se concentrations in sediments after microbial populations were stimulated by the addition of nutrients and/or organic materials and the maintenance of ideal growing conditions (moisture, aeration). It should be noted that such remediation processes are not appropriate for Hg because the methylated Hg compounds are very toxic, unlike dimethylselenide.

Another experimental biological remediation method that is being studied involves the use of hyperaccumulator plant species that accumulate metals in their tissue to very high concentrations. *Thlaspi alpestre* is a Zn hyperaccumulator that has been found growing when soil Zn concentrations were extremely high. Biomass production by *Thlaspi* is very low, and it is hoped that this can be improved — with biotechnology — to the point that Zn removal from soils with such a hyperaccumulator might be feasible.

Figure 7-7 illustrates the effectiveness of various soil treatments in reducing the bioavailability of Zn in a metal-contaminated alluvial soil. In this situation, the indicator of bioavailable Zn was that extracted from the soil with a 0.5 M KNO$_3$ solution. The treatments included the addition of lime, P, cattle manure, sewage sludge, poultry litter, or various combinations of lime and cattle manure. The treatments produced KNO$_3$ extractable Zn concentrations ranging from 3.7 to 63.3 mg/kg

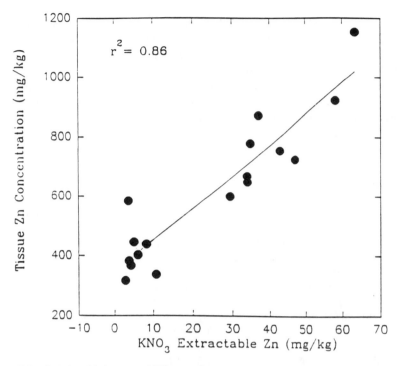

$r^2 = 0.86$

Figure 7-7 Relationship between KNO_3-extractable Zn and soybean tissue Zn concentration after various amendments were made to a metal-contaminated alluvial soil in an effort to reduce Zn bioavailability. (Data from Pierzynski and Schwab, 1993.)

with a corresponding range of soybean tissue Zn concentrations of 318–1153 mg/kg. The manipulation of the bioavailable Zn levels was done without changing the total Zn concentration of the soil.

It should be noted that phytoxicities can protect the human food chain. This phenomenon is called the *soil-plant barrier,* and it refers to the situation where a plant experiences phytoxicity at a trace element concentration below that which would be harmful for humans to consume the plant as food. This is true for Fe, Pb, Hg, Al, Ti, Ag, Au, Sn, Si, and Zr but not for Zn, Cd, Mn, Mo, Se, or B. The mechanisms responsible for this include insolubility or strong retention of the element in the soil that prevents plant uptake (e.g., Cd versus Pb), immobility of the element in nonedible portions of plants that prevents movement into edible portions (e.g., roots versus aboveground portions), or phytotoxicity that occurs at concentrations in the edible portions of plants below that harmful to animals or humans. Note that direct ingestion of contaminated soil or dust bypasses the soil-plant barrier.

7.5 ADDITIONAL REMARKS

High doses of certain trace elements can have detrimental effects on plants, animals, humans, and aquatic organisms. Contamination of agricultural soils, however, is not widespread. This factor helps to protect the human food chain since most individuals consume food grown across a wide geographic area. A general description of individuals at greatest risk for unacceptable doses of trace elements would be those who reside in areas with trace element contaminated soils; are exposed to dust, etc.; and have a large proportion of their diets comprised of locally grown produce, meat, and milk. The soil-plant barrier may also provide additional protection for the human food chain. In a relative sense, animals may be at greater risk since they spend their lives in a small area. Concern for trace elements in soils focuses on reductions in soil quality from the viewpoint of unacceptable concentrations of pollutants and limitations in soil function.

REFERENCES

Bowie, S. H. U. and Thornton, I., Ed., *Environmental Geochemistry and Health,* Kluwer Academic Publ., Hingham, MA, 1985.

Kim, S. J., Chang, A. C., Page, A. L., and Warneke, J. E., Relative concentrations of cadmium and zinc in tissue of selected food plants grown on sludge-treated soils, *J. Environ. Qual.,* 17, 568, 1988.

Lexmond, T. M. and deHaan, F. A. M., in Proc. Int. Semin. Soil Environ. Fertil. Manage. Intensive Agric., Tokyo, Soc. Sci. Soil Manure, Japan, 1977, 383.

Minnich, M. M., McBride, M. B., and Chaney, R. L., Copper activity in soil solution. II. Relation to copper accumulation in young snapbean plants, *Soil Sci. Soc. Am. J.,* 51, 573, 1987.

Nriagu, A silent epidemic of environmental metal poisoning, *Environ. Pollution,* 50, 139, 1988.

Page, A. L., personal communication, 1992.

Pierzynski, G. M. and Schwab, A. P., Bioavailability of zinc, cadmium, and lead in a metal contaminated alluvial soil, *J. Environ. Qual.,* 22, 247, 1993.

Pierzynski, G. M., Agronomic Considerations for the Application of a Molybdenum-Rich Sewage Sludge to an Agricultural Soil, M. S. thesis, Michigan State University, East Lansing, MI, 1985.

Pouyat, R. V. and McDonnell, M. J., Heavy metal accumulations in forest soils along an urban-rural gradient in southeastern New York, U.S.A., *Water, Air, Soil Pollut.,* 57–58, 797, 1991.

Tsuchiya, K. Environmental Pollution and Health Effects, in *Cadmium Studies in Japan: a Review,* Tsuchiya, K., Ed., Elsevier/North Holland Biomedical Press, New York, 1978, 144.

SUPPLEMENTARY READING

Three excellent references for trace elements in soils are: Adriano, D. C., *Trace Elements in the Terrestrial Environment,* Springer-Verlag, New York, 1986; Kabata-Pendias, A. and Pendias, H., *Trace Elements in Soils and Plants,,* CRC Press, Boca Raton, FL, 1992; and Adriano, D. C., Ed., *Biogeochemistry of Trace Metals,* CRC Press, Boca Raton, FL, 1992.

Bar-Yosef, et al., Eds., *Inorganic Contaminants in the Vadose Zone,* Springer-Verlag, New York, 1989.

Jacobs, L. W., Ed., Selenium in Agriculture and the Environment, Soil Science Society of America, Special Publ. no. 23, American Society of Agronomy, Madison, WI, 1989.

8 ORGANIC CHEMICALS IN THE ENVIRONMENT

8.1 INTRODUCTION

The production of synthetic organic chemicals has increased rapidly since the turn of the century, with the growth in the production industry resulting in new and improved materials that have affected the way we live. However, with the good comes the bad; increased production (about 60 million metric tons are manufactured yearly in the United States) and utilization of synthetic organic compounds (such as pesticides, lubricants, solvents, fuels and propellants) have increased the number of incidents whereby organic chemicals have accidently entered into atmosphere, hydrosphere, and soil environments. Humans contribute enormous amounts of organic chemicals to the environment every day. Petroleum products are the dominant source of hydrocarbons released into the atmosphere due to human activities with nearly 50% related to gasoline emissions.

Organic chemicals are also present in surface- and groundwaters. Groundwaters, once thought to be pristine, have been found to contain a broad spectrum of organic chemicals that are both indigenous and anthropogenic in nature. In a study of groundwaters commissioned by the U.S. Congress (Office of Technology Assessment, 1984), there were approximately 175 different natural and man-made organic chemicals identified. Many of these natural organic chemicals, as well as microorganisms present in surface- and groundwaters will react with chlorine, which is commonly used in drinking water treatment, to produce trihalomethanes, a group of volatile halogenated organic compounds.

In the United States alone there are approximately one half million metric tons of pesticides used annually, most of which are used in crop production to control weeds, insects, and diseases. An estimated 25,000 metric tons of pesticides are also used in nonagricultural situations such as pest control in lawns, flower beds, golf courses, forests, and in and along waterway, utility, and rail easements. In a survey of pesticides in groundwater in the U.S. (Parsons and Witt, 1988), 67 different pesticides were found in 33 states (a total of 35 states participated in the survey); however, only 17 pesticides in 17 states were detected at levels greater than the Environmental Protection Agency (EPA) Health Advisory (HA) level. The percentage of groundwaters exceeding the HA level for any particular pesticide ranged from 0 to 10.1%.

The focus of this chapter is on organic chemicals in soil and aquatic environments with particular emphasis on pesticides. Specific areas that will be covered include:

- adverse effects of organic chemicals on organisms and human health
- fate of pesticides and organic chemicals in the environment
- remediation strategies for contaminated aquifers and soils
- pest management

8.1.1 Sources of Organic Chemicals

The introduction of organic chemicals into the environment can occur by design, accident, or neglect. There are a variety of sources that can contribute to the release of organic chemicals into the environment; industrial sites typically manufacture, store, and distribute the greatest amount of organic chemicals and therefore have a higher probability of contaminating the environment. Several industrial sites that have the potential for contaminating surface- and groundwaters and soils are listed in Table 8-1. In addition, other sources that may be a cause for concern are land application of wastes, feedlot operations, landfills, to name a few. Although the quality of groundwater is influenced by natural processes, as well as human activities, organic chemical contamination is generally a consequence of improper management by humans.

8.1.2 Categories of Pesticides

Pesticides are classified according to the target organisms they are designed to control (Table 8-2). Of the target organisms, weeds by far cause the greatest economic loss due to their interference in crop production. It is not surprising, therefore, that herbicides are the most commonly used pesticides, comprising over 60% of the pesticides applied in the U.S. every year. Insects are perceived as our next greatest problem, with insecticide use totaling approximately 25%. There is a great deal of indiscriminate use of insecticides in homes and buildings, which has prompted some individuals to suggest that a considerable amount of money is needlessly wasted. Fungicides are the third most used pesticides, totaling about 6%.

Table 8-1 Some Examples of Industrial Sites That Have the Potential for Contaminating the Environment

Acid/alkali plants	Pharmaceutical, perfumes,
Asbestos	cosmetics, toiletries
Chemical and allied products	Polymers and coatings
Explosives and munitions	Railway yards
Gas works and storage	Scrap yards
Metal treatment and finishing	Shipping docks
Mining and extractive industries	Smelting and refining
Oil production	Sewage treatment
Oil storage	Tanning
Paints	Waste disposal
Pesticide manufacture	Wood preservation

Table 8-2 Categories of Pesticides Used to Control Unwanted Pests

Pesticide	Control
Herbicide	Prevents the growth of weeds in agricultural crops, lawns, golf courses, etc. or applied directly to weeds that are established
Insecticide	Kills or controls harmful or undesirable insects that live on plants, in animals, or in buildings
Fungicide	Protects plants from infestations by diseases; usually used prior to conditions which are favorable to disease development
Bactericide	Controls bacteria that can cause damage to fruit and develop galls on plants
Nematicide	Protects young plant roots from microscopic soil worms that infect plants and feed on their roots
Acaricide	Controls mites and spiders that can infest or damage agricultural crops or ornamental plants
Rodenticide	Kills mice and rats living in homes and other buildings and prevents infestations or losses of food products in storage

8.2 ADVERSE EFFECTS

The potential for negative environmental impacts due to the misuse or accidents involving organic chemicals is a concern because of the difficulty in many cleanup procedures and because of the direct and indirect consequences of misuse and accidents on surrounding ecosystems. Care must be taken in the manufacturing, storage, transportation, and handling of organic chemicals. With adequate training and precautions, many of the incidences of contamination that occur could be avoided.

Several noteworthy examples of adverse effects due to pesticide misuse have been detailed in scientific literature. Many studies have cited the carryover of residual levels of pesticides in soils and aquatic systems as resulting in pesticide toxicity to plants and fish. Bioaccumulation is also of concern since over time low levels of pesticides absorbed or ingested can accumulate to toxic levels in the organism itself, its offspring, or organisms higher on the food chain. Many of the pesticides are toxic to organisms other than those specifically targeted. At high enough levels, most of the pesticides in use can also be toxic to humans. One of the greatest concerns with the development/use of pesticides is their slow breakdown and ability to accumulate in organisms. However, it should be stated clearly that pesticides, if used properly, can increase food production or control insects that may have a devastating effect on the quality of animal and human life.

8.2.1 Mode of Biological Action of Pesticides

The mode of action of pesticides refers to the mechanism by which the pesticide kills or interacts with the target organisms. Based on their modes of biological action, pesticides are generally classified as either *contact* or *systemic*. Contact pesticides kill target organisms by weakening or disrupting cellular membranes which, in turn, results in the loss of cellular constituents. If the contact pesticide has an acute reaction with the target organism, death may be extremely rapid. Systemic pesticides must be absorbed or ingested by the target organism so that it may interfere with the organism's physiological (e.g., cell division, chlorophyll formation, tissue development) or metabolic (e.g., respiration, enzyme activity, photosynthesis) processes. Systemic pesticides are generally slow acting and may require days, weeks, or longer periods of time before results become evident.

8.2.2 Impacts on Microorganisms

Organic chemicals introduced into the environment can have a devastating effect on certain organisms, which may or may not be the intended purpose of the organic chemical. Soil and aquatic ecosystems contain a multitude of microorganisms, many of which are beneficial. Contamination of these ecosystems by organic chemicals, such as when high concentrations of pesticides are inadvertently used or when surface runoff containing pesticides is introduced into surface waters, can result in a reduction in microbial activity. However, in some situations enhancement of microbial activity may occur.

The theoretical response of soil microorganisms to different herbicides is shown in Figure 8-1, which could also relate to other organic chemicals besides herbicides. "Initial application" in Figure 8-1 represents the first time the herbicide was used. "Subsequent applications" refers to the response of the ecological system to continued application of the herbicide. The initial herbicide effects can be summarized as follows: herbicide A — no inhibition of soil microorganisms, rapid degradation rate, increase in microbial populations; herbicide B — no inhibition of soil microorganisms, moderate degradation rate, increase in microbial populations; herbicide C - kills sensitive soil microorganisms, increase in herbicide-degrading microorganisms over time, slow degradation rate; herbicide D — initially herbicide kills sensitive microorganisms, herbicide metabolites build up and inhibit certain microorganisms, resistant microorganisms eventually degrade herbicide; herbicide E — kills sensitive microorganisms, resistant microorganisms increase but do not degrade herbicide. On further herbicide application, herbicides A, B, and C are rapidly degraded; herbicide D and its metabolites are degraded at a faster rate; and herbicide E concentrations build up due to lack of degrading microorganisms. These are just a few examples of how soil ecosystems might respond to organic chemical inputs. In an extreme case of contamination, the number and activity of soil microorganisms may be reduced to essentially zero.

8.2.3 Impacts on Plants

Plants that are sensitive to pesticides can show rapid signs of growth irregularly, loss in biomass, or death. As was discussed in the last section for microorganisms (Figure 8-1), plants also are capable of building up a resistance to certain pesticides. Higher concentrations of the pesticide would then be required to achieve a similar level of control. Pesticides other than herbicides, such as insecticides, can also affect plants that are not the specific target organism for which the pesticide was designed to control. Table 8-3 lists effects that certain herbicides and insecticides can have on nontargeted plants located within soil and water environments.

8.2.4 Human Health

Pesticides are used to benefit mankind by controlling diseases such as malaria and typhus, weed infestation that reduces crop productivity, and insects outbreaks.

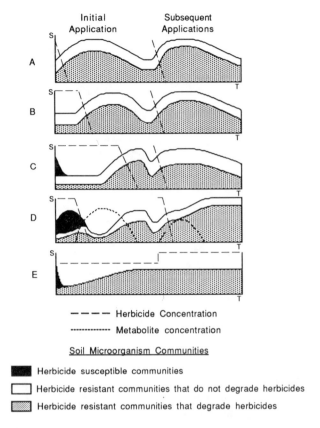

Figure 8-1 Theoretical response of soil microbial communities and degradation rates of herbicides (A-E) over time. Time is represented on the abscissa and herbicide concentration on the ordinate axis. (Redrawn from Cullimore, 1971.)

However, high doses of some pesticides can be harmful to humans. Laboratory research has shown that high doses of certain pesticides given to animals can cause cancer, mutagenesis, neuropathy, and even death. The pesticide concentrations given in these experiments are generally well above levels that are recommended by the manufacturers for the intended purpose of the pesticide. Although the exposure of humans to pesticides is relatively low (except in cases of misuse or accidents), we must continue to be aware of the potential impacts of pesticides. Therefore, it is necessary for anyone working with pesticides to follow directions and take every precaution to protect themselves other humans and our environment from becoming contaminated.

8.2.5 Groundwater Contamination

Groundwater is used for several purposes; potable water for greater than 50% of the U.S. population and for approximately 97% of rural residents comes from groundwater sources, and about 40% of agricultural irrigation water is also derived from groundwater sources. Groundwater quality is an issue that has generated much debate and concern. Several definitions have been proposed to define groundwater contamination,

Table 8-3   Effects of Some Herbicides and Insecticides on Nontargeted Plants in Soil and Water Environments

Pesticide type	Location	Effect
Herbicides		
Aromatic acid	Soil	Carryover in residue effects subsequent crop; soluble herbicides (e.g., picloram) can injure adjacent plants
	Water	Kills or inhibits some aquatic plants
Amides, analines, nitriles, esters, carbamates	Soil	Some persistence resulting in residue carryover affecting subsequent crop
	Water	Surface erosion may transport herbicide to water bodies
Insecticides		
Organochlorine	Soil	Residue carryover can affect subsequent crops; transport to surface waters may affect aquatic plants
	Water	Contaminated waters can affect plants if waters used for irrigation
Organophosphates, carbonates, pyrethroid	Soil	Usually short lived, thus little effect on plants
	Water	Toxic to certain algae

Source: Madhun and Freed, 1990.

but the definition by Miller (1980) which states, "Groundwater contamination is the degradation of the natural quality of groundwater as a result of man's activities," places the blame completely and clearly on humans and their misuse of the environment. Contamination of groundwaters is receiving much attention; however, assessing the magnitude of the problem will be difficult unless the contaminant sources can be identified and prevented from further contaminating the environment. The Office of Technology Assessment identified 33 principal sources that have the potential for groundwater contamination, and grouped them into six categories. Table 8-4 lists the six categories of contaminant sources which are grouped based on the general nature of the contaminant activity.

8.3 PREDICTING THE FATE OF PESTICIDES AND OTHER ORGANIC CHEMICALS

Reliable predictions of the movement and fate of organic chemicals in the environment (Figure 8-2) is important for determining the impact organic chemicals have on our environment. For pesticides, their proper use requires a knowledge of factors that can influence their movement and fate, as well as human safety, environmental quality, and pesticide efficacy. Surrounding ecosystems can be impacted if conditions are conducive to pesticide drift, leaching, or surface runoff. Table 8-5 outlines some of the factors related to pesticide use that can impact environmental quality and pesticide effectiveness.

The potential movement or transfer of organic chemicals, and their ultimate degradation after being introduced into the soil environment, are determined by chemical, physical, and biological processes. Each organic chemical has its own molecular structure which determines, at least partially, the degree to which it will

Table 8-4   Sources of Organic Chemicals That Have the Potential for Contaminating Groundwater and Soil

Category	Sources	Description
I	Intentional discharge	Septic tanks and cesspools
		Injection wells (hazardous wastes)
		Land application (sludge, wastewaters hazardous wastes)
II	Containment	Landfills and open dumps
		Surface impoundments
		Waste tailings and piles
		Animals burial sites
		Underground storage tanks
		Storage containers
III	Transportation	Pipelines
		Rail and truck
IV	Discharge from planned activities	Pesticide application
		Animal feed operations
		Urban runoff
		Atmospheric pollutants
V	Induced flow	Production wells (oil and gas)
		Construction excavation
VI	Natural	Groundwater-surface water interactions
		Wetlands
		Underground deposits (coal, oil, gas)

Source: Office of Technology Assessment, 1984.

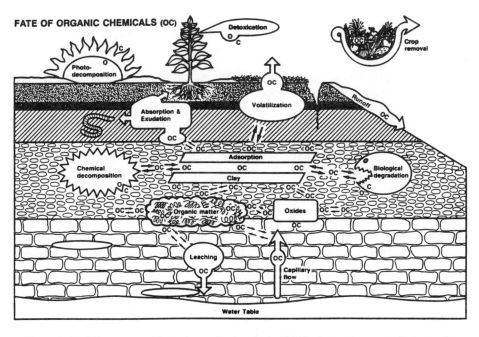

Figure 8-2   Processes and fate of organic chemicals (OC) in the environment. Both transfer and degradation processes are depicted in the figure; transfer processes have the organic chemical intact (i.e., OC) whereas in degradation processes the OC is separated (i.e., O/C). (Adapted from Weber and Miller, 1989. With permission.)

Table 8-5 Impacts of Pesticide Management Decisions on Both Environmental Quality and Pest Control Efficacy

Management decision	Impacts	
	Environmental quality	Pest control
Type of pesticide	High sorption — low leaching	Efficacy of control dependent on pesticide
	Fast degradation — not transported Toxicity — determines maximum allowable concentration in water	
Type of application	Soil incorporated — greater chance for groundwater contamination	Toxicity to target pest enhanced by proximity of pesticide to target
	Ground application — decreases risk of pesticide drift	Wet soil conditions may hinder optimum time of applicant
Spray technology	Large droplets — reduced physical drift	Large drops may decrease efficacy
Timing of application	Close to rainfall or irrigation — increases chance for groundwater contamination	Efficacy of pesticide depends on proper timing
Irrigation schedule	Planned irrigation — reduces potential for ground- or surface water contamination	Changes in soil moisture may affect pests
Tillage practice	Minimum tillage effects — reduces runoff, requires increased levels of pesticide, greater risk of pesticide leaching or drift	Increases pest densities which requires additional need for pesticides
Proper application procedures	Pesticide spills or misuse — increases water and soil contamination	No negative effect on pest control

Source: Shoemaker, 1989.

react in the environment. Some organic chemicals contain charged functional group sites that enhance the chance of them being adsorbed to soils; positively charged organic chemicals are adsorbed to the negatively charged sites on soil clays and organic matter (see Chapter 2 for more information on soil properties). Other organic chemicals may persist (e.g., recalcitrant organic compounds) in soils for long periods of time if chemical and microbial transformations are suppressed due to the toxicity of the organic chemical under certain environmental conditions.

Different transfer and degradation processes controlling the movement and fate of organic chemicals are listed in Table 8-6. Physical drift is generally a concern during the application of pesticides, but could also occur at times of high winds or during rainstorms when pesticides are washed off vegetation. Small droplets have a greater potential for being transported by wind action than do large droplets. Other processes generally relate to what occurs to the organic chemicals once they are introduced into soil environments. Within the next sections, we will discuss many of these processes and other characteristics of organic chemicals that influence their fate and transport.

8.3.1 Plant Uptake

Organic chemicals, such as herbicides, are absorbed by plants either through their roots or aboveground foliage. Uptake is dependent on the organic chemical and the

Table 8-6   Movement and Fate of Organic Chemicals in the Environment, with Particular Reference to Pesticides

Process	Consequence	Factors
Transfer (processes that relocate organic chemicals without altering their structure)		
Physical drift	Movement of organic chemical due to wind action	Wind speed, size of droplet, distance to physical object
Volatilization	Loss of organic chemical due to evaporation from soil, plant, or aquatic ecosystems	Vapor pressure, wind speed, temperature
Adsorption	Removal of organic chemical by interacting with plants, soils, and sediments	Clay and organic matter content, clay type, moisture
Absorption	Uptake of organic chemical by plant roots or animal ingestion	Cell membrane transport, contact time and susceptibility
Leaching	Translocation of organic chemical either laterally or downward through soils	Water content, macropores, soil texture, clay, and organic matter content
Erosion	Movement of organic chemical by water or wind action	Rainfall, wind speed, size of clay and organic matter particles with absorbed organic chemicals on them
Degradation (processes that alter the chemical structure)		
Photochemical	Breakdown of organic chemicals due to the adsorption of sunlight (i.e., ultraviolet light)	Structure of organic chemical, intensity and duration of sunlight, exposure
Microbial	Degradation of organic chemicals by microorganisms (see Chapter 2 for discussion on soil microorganisms)	Environmental factors (pH, moisture, temperature) nutrient status, organic matter content
Chemical	Alteration of organic chemical by chemical processes such as hydrolysis and redox reactions	High and low pH, same factors as for microbial degradation
Metabolism	Chemical transformation of organic chemical after being absorbed by plants or animals	Ability to be absorbed, organism metabolism, interactions within the organism

Source: Marathon-Agricultural and Environmental Consulting, Video Cassettes — *Fate of Pesticides in the Environment,* 1992.

plant species. Seeds are also capable of adsorbing organic chemicals, which can occur either before or after seed germination. Adsorption of organic chemicals on the outer surface of seeds increases the chance of absorption, which can take place by nonmetabolic processes when seeds are imbibing water or through diffusion processes. Factors that influence seed uptake of organic chemicals are related to the properties of the chemical (concentration, structure, solubility, and diffusion rate), soil (temperature and pH), and seed (size, characteristics, and permeability of the seed coat).

Plants are also capable of absorbing organic chemicals through their aboveground parts, which include stems, buds, and leaves. Succulent and perennial woody plants are noted for their uptake of herbicides through their stems. Buds are the primary target of contact herbicides since entry of the herbicide into the bud generally guarantees plant kill. Leaves absorb organic chemicals from both the upper and lower leaf surfaces; however, absorption is often faster through the lower surfaces because it has a thinner cuticle. Nonpolar organic chemicals enter into leaves more readily than do polar organic chemicals, or many inorganic constituents. The mechanism of entry for oily and aqueous solutions is thought to follow different pathways.

Many organic chemicals can be readily taken up by roots and foliage and are transported through plants within two systems — the symplast (living plant tissue) and apoplast (nonliving plant tissue). Organic chemicals entering the symplast system are exposed to many enzymes which are capable of interacting with the organic chemical. However, the symplast is enclosed by the apoplast thus preventing direct contact between organic chemicals and the symplast living tissue. The apoplast system consists of cell walls and xylem that form an interconnected continuum throughout the plant. It is within the apoplast that toxic pesticides and other organic chemicals can be translocated over short and long distances in plants.

Once organic chemicals are absorbed by plants, several reactions can occur that transform the organic chemical. For example, plants which are capable of adsorbing herbicides may be able to transform, or metabolize, them to nonphytotoxic levels by one or more of the following reactions: *oxidation-reduction, hydrolysis, hydroxylation, dehalogenation, dealkylation, conjugation,* or *β-oxidation.* Each of these reactions alters the herbicide structurally so that its chemical nature is different. Although plants are similar to microorganisms, insects, and mammals in their capability to metabolize organic chemicals, rates for plant transformations are generally slow.

8.3.2 Solubility

The solubility of an organic chemical is important to its fate and mobility, because highly soluble chemicals tend to be rapidly distributed within soil and hydrosphere environments. *Aqueous solubility* of organic chemicals is determined based on the total amount that dissolves in pure water at a specified temperature. When the aqueous solubility of an organic chemical is exceeded, a solid or organic phase will exist in addition to the aqueous solution. Solubilities of common organic chemicals generally fall in the range of 1–100,000 mg/kg (weight of organic chemical per weight of pure water); however, many have higher solubilities. The difference between the least soluble and most soluble of the organic chemicals is approximately one billion (see Figure 8-3 for the range in solubilities).

Solubility of organic chemicals in water is a function of temperature, pH, ionic strength (concentration of soluble salts), and other organic chemicals such as dissolved organic carbon (DOC). Most organic chemicals become more soluble with increasing temperature, although the reverse is true for some. The solubility of certain types of organic chemicals such as organic acids generally increases with increasing pH, whereas organic bases are expected to behave in an opposite manner. Soluble salts will commonly reduce organic solubility, which explains why organic chemicals in oceans tend to be less soluble than in fresh waters. Several studies have indicated there is usually an increase in the solubility of many poorly water soluble organic chemicals with higher levels of DOC. Evidently there is an interaction between the DOC and organic chemicals that enhances its apparent solubility.

Various methods are used for the estimation of organic chemical solubility. Two frequently used methods are based on (1) chemical structure and (2) octanol/water partition coefficients. The former method has been developed for estimating the

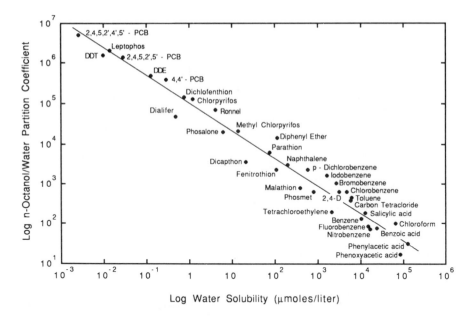

Log Water Solubility (μmoles/liter)

Figure 8-3   Relationship between octanol-water partition coefficients (K_{ow}) and solubility of several organic chemicals. Note the extensive range in solubilities of the organic chemicals. (Adapted from Chiou et al., 1977. With permission.)

aqueous solubility of particular groups of compounds, for example, with aliphatic and aromatic hydrocarbons. A significant amount of data is available on the relationship between aqueous solubility and octanol/water partition coefficients. *Octanol/water coefficients* (K_{ow}) are determined by the following equation:

$$K_{ow} = \frac{\text{Conc. of organic chemical in octanol (mg)}}{\text{Conc. of organic chemical in water (mg)}} \tag{8-1}$$

The relationship between water solubility and K_{ow} is shown in Figure 8-3. As seen on this figure, there is an inverse relationship between solubility and K_{ow}.

8.3.3 Half-life

The *half-life* $(t_{1/2})$ of a reaction refers to the amount of time required for half of the reactant to be converted into a product or when the reactant concentration is half of its initial level. For organic chemicals, half-lives can be calculated for different types of reactions such as volatilization, photolysis (decomposition by sunlight), leaching potential (adsorption-desorption characteristics), and degradation (chemical and microbial). Half-life values are important for understanding the potential environmental impact of a particular organic chemical. For instance, if a highly toxic organic contaminant is accidently spilled into a lake and the rate of photolysis is rapid (suggesting a small $t_{1/2}$), the consequences may be minimal if the photolysis products are harmless. However, if a moderately toxic contaminant is spilled and it has a very

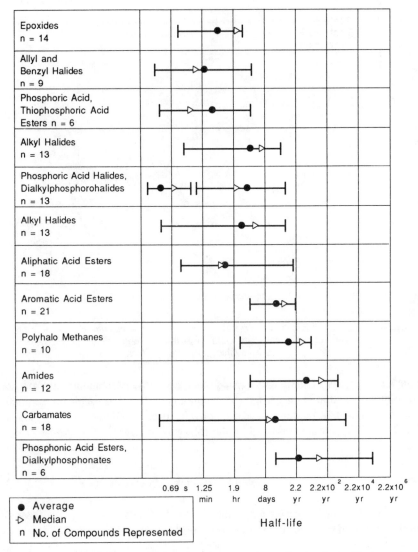

Figure 8-4   Hydrolysis half-lives $(t_{1/2})$ for several groups of organic chemicals. (Redrawn from Harris, 1982a. Data from Mabey and Mills, 1978.)

slow rate of photolysis (indicating a large $t_{1/2}$), then the environmental impact may be substantial. This kind of scenario can also be used as an example for determining environmental impacts related to volatilization, leaching potential, and degradation characteristics of various organic chemicals.

The ranges in half-lives for the hydrolysis of several groups of organic chemicals are shown in Figure 8-4. Hydrolysis reactions involve the chemical transformation between an organic chemical and water which results in the breaking of one bond while forming a new carbon-oxygen bond (e.g., $C–X + H_2O \rightarrow C–OH + H^+ + X^-$). Hydrolysis is considered one of the most important reactions that determine the fate

**Table 8-7 Examples of Organic Functional Groups
That Are Either Resistant or Potentially Susceptible
to Hydrolysis**

Resistant	Susceptible
Alkanes, alkenes, alkynes	Alkyl halides
Benzene/biphenyls	Amides
Polyaromatic hydrocarbon (PAH)	Amines
Halogenated aromatics/PCBs	Carbamates
Dieldrin/aldrin and related	Carboxylic acid esters
halogenated hydrocarbon pesticides	Epoxides
Aromatic amines	Nitriles
Alcohols	Phosphonic acid esters
Phenols	Phosphoric acid esters
Glycol	Sulfonic acid esters
Ethers	Sulfuric acid esters
Aldehydes	
Ketones	
Carboxylic acids	

Source: Harris, 1982a.

of organic chemicals in aquatic environments. Not all organic chemicals can undergo hydrolysis, since many do not possess functional groups susceptible to hydrolysis reactions (Table 8-7). Hydrolysis reactions are also generally pH dependent.

8.3.4 Volatilization

Volatilization of natural and synthetic organic chemicals is responsible for the transfer of organic chemicals from aquatic and soil environments into the atmosphere. Information on the rate at which organic chemicals volatilize is important for understanding their persistence in the environment.

Factors that affect the volatilization of contaminants from waters are dependent on the chemical and physical properties of the organic chemical (e.g., solubility and vapor pressure), interactions with suspended materials and sediment, physical properties of the water body (depth, velocity, and turbulence), and properties of the water-atmosphere interface. Rates of volatilization, expressed as a half-life, can vary from hours to years or more. Half-lives for the volatilization of trichloroethylene (TCE) and the pesticide dieldrin from water have been estimated at 3–5 hr for TCE and close to a year for dieldrin. Table 8-8 lists some additional organic chemicals and their potential volatilization from water.

Factors that influence the volatility of organic chemicals from soils include the chemical's intrinsic physiochemical properties (i.e., vapor pressure, solubility, structure and nature of functional groups, and adsorption-desorption characteristics), concentration, soil properties (soil moisture content, porosity, density, and organic matter and clay contents); and environmental factors (temperature, humidity, and wind speed). The initial step in the volatilization of organic chemicals from soils is the ability to evaporate, which represents a change from a solid or liquid to a vapor. After evaporation, vapor moves through the soil and disperses into the atmosphere by diffusion or turbulence. Laboratory and field measurements of volatilization half-lives for several pesticides range from 0.7–3.0 days for lindane to 42 and 45 days for DDT and atrazine, respectively.

Table 8-8   Examples of Organic Chemicals and
Their Potential Volatility from Water

Volatility potential	Organic chemical	Half-life $(t_{1/2})$
Low	Dieldrin	327 d
	3-Bromo-1-propanol	390 d
Medium	Penanthrene	31 hr
	Pentachloraphenol	17 d
	DDT	45 hr
	Aldrin	68 hr
	Lindane	115 d
High	Benzene	2.7 hr
	Toluene	2.9 hr
	O-xylene	3.2 hr
	Carbon tetrachloride	3.7 hr
	Biphenyl	4.3 hr
	Trichlorethylene	3.4 hr

Source: Thomas, 1982.
Note: Half-lives given in days (d) or hours (hr).

8.3.5 Photolysis

Photochemical reactions involving sunlight are extremely important in determining the fate of contaminants in aquatic environments, and may also play a role in the degradation of organic chemicals at soil surfaces. In aquatic environments, photolysis can occur by either direct or indirect processes. In direct photolysis, sunlight is absorbed directly by the organic chemical resulting in a chemical transformation. The rate of direct photolysis is dependent on sunlight intensity and overlapping spectral characteristics of solar radiation and the organic chemical. With indirect photolysis, other substances such as DOC, clay minerals, or inorganic elemental species absorb sunlight and either initiate a series of reactions that ultimately transform the organic chemical or transfer the excitation energy to the organic chemical.

Atmospheric ozone absorbs solar radiation below wavelengths of 290 nm. Therefore, direct photolysis by sunlight will not occur if the organic chemical in question does not absorb radiation at wavelengths above 290 nm. The intensity of sunlight that reaches the surface of the earth is determined by the thickness of the atmosphere and the angle of incident radiation, which is dependent on latitude, season, and time of day. Sunlight intensity is greatest in summer and least in winter.

Half-lives of organic chemicals that undergo photolysis in aqueous environments are generally in the range of hours to months. Photolysis half-life values for pesticides, polycyclic aromatic hydrocarbons, and some miscellaneous compounds are listed in Table 8-9. Unlike aqueous environments, photolysis reactions in soils are difficult to determine because of the heterogeneous nature of soils and the lack of sunlight penetration. Photolysis is primarily a surface phenomenon that is prevented if the chemical is incorporated into the soil.

8.3.6 Sorption-Desorption

Sorption-desorption behavior of organic chemicals is conceivably the most important process affecting organic contaminants. Understanding the interactions that occur between soils or sediments and an organic contaminant once they enter soil and

Table 8-9   Half-Life Values for Several Organic Chemicals That
Undergo Direct Photolysis

Class	Organic chemical	Half-life $(t_{1/2})$
Pesticides	Trifluralin	~1 hr
	Malathion	15 hr
	Carbaryl	50 hr
	Sevin	11 d
	Methoxychlor	29 d
	2,4-D, methyl ester	62 d
	Mirex	1 yr
Polycyclic aromatic hydrocarbons	Pyrene	0.7 hr
	Benz[a]anthracene	3.3 hr
	Phenanthrene	8.4 hr
	Fluoranthene	21 hr
	Naphthalene	70 hr
Miscellaneous	Benz[f]quinoline	1 hr
	Quinoline	5–21 d
	p-Cresol	35 d

Source: Harris, 1982b.

aquatic environments is important in determining the contaminant behavior, and therefore its movement and fate in the environment. Predictive modeling of contaminant fate and transport requires reliable information on the sorption/desorption behavior of contaminants under variable conditions. Sorption of organic chemicals by clay and organic matter materials occurs by one or more of the following interactions: van der Waals forces, H bonding, dipole-dipole interaction, ion exchange, covalent bonding, protonation, ligand exchange, cation bridging, water bridging, and/or hydrophobic partitioning.

Sorption of many contaminants by soils and sediments has been shown to be an effective means of reducing their mobility. Sorption can also affect the bioactivity, persistence, biodegradability, leachability, and volatility of organic chemicals. The type and nature of functional groups on an organic chemical largely determine its ability to be adsorbed. In soils, clay and metal oxide surfaces and organic matter are the dominant materials responsible for the sorption of organic contaminants. Clay surfaces can act as a source of adsorption sites for one contaminant that in turn can influence the adsorption of other contaminants. An example of this is discussed in Section 8.4.4 as a method for the containment of contaminants.

Modeling the sorption of organic chemicals by soils is frequently done by using adsorption isotherms. Sorption data are most commonly described by using either the Freundlich or the Langmuir equation, similar to the use discussed for P in Chapter 5.

The Freundlich adsorption equation is:

$$x/m = KC^{1/n} \tag{8-2}$$

where

x/m = mass of organic chemical adsorbed per unit weight of soil
K and n = empirical constants
C = equilibrium concentration of the organic chemical

The value of K is a measure of the extent of sorption. The linear form of the Freundlich equation is obtained by logarithmic transformation:

$$\log (x/m) = \log K + 1/n \log C \qquad (8\text{-}3)$$

A plot of log (x/m) vs log C should produce a straight line, with 1/n equal to the slope and log K the intercept.

The Langmuir adsorption equation is:

$$x/m = KbC/(1 + KC) \qquad (8\text{-}4)$$

where

    x/m and C = same as described above,
    K   = adsorption constant that is related to binding strength
    b   = maximum amount of organic chemical that can be sorbed by
                the soil

The linear form of the Langmuir equation is:

$$C/(x/m) = 1/Kb + C/b \qquad (8\text{-}5)$$

If a plot of C/(x/m) vs C is a straight line, then the adsorption data conform to the Langmuir equation, and b can be calculated from the slope and K from the intercept. Equations 8-2 and 8-4 are the general forms of the Freundlich and Langmuir equations, respectively. Equations 5-1 and 5-2 are the same except the variables have been written specifically for P.

For many organic chemicals, and especially nonpolar types, sorption constant can be calculated that is relatively constant, independent of soil type, and specific for the organic chemical used. The constant is calculated using the Freundlich K and the percent soil organic carbon (% OC).

$$K_{oc} = (K/\%OC) \times 100 \qquad (8\text{-}6)$$

The K_{oc} is essentially a coefficient that describes the distribution of the organic chemical between aqueous and soil organic matter phases. Being relatively constant, K_{oc} values are often used to predict K values for soils with known organic carbon contents. The K value in turn defines the distribution of the contaminant between soil and water.

8.3.7 Abiotic and Biotic Transformations

Both abiotic and biotic reactions, alone or in combination, are responsible for the transformations of organic chemicals in soil and aquatic environments. Under certain conditions abiotic reactions may dominate, whereas under other conditions biotic reactions may prevail. Degradation of organic chemicals is often assumed to occur by biotic processes; however, abiotic reactions may occur simultaneously. Many organic chemical transformations are mediated by microorganisms, but the actual reaction is an abiotic process.

The principal abiotic transformation reactions that occur in aquatic environments include *hydrolysis, oxidation-reduction* (redox), and *photolysis;* in sediments, hydrolysis and redox reactions dominate. Oxidation reactions that take place in aquatic environments can be mediated by direct or indirect photolysis reactions, which depend on the organic chemical and substrates present. Nonphotolytic oxidation can occur directly by ozone, or by catalytic properties of certain metals. Abiotic reduction of organic chemicals may also be catalyzed by certain metal species, with Fe and Mn being the most important. Redox reactions that occur in sediments may follow a similar route as shown for soils.

In soils, abiotic transformations take place in the liquid phase (i.e., soil solution) and at the solid-liquid interface. In the soil solution, hydrolysis and redox reactions are the most common abiotic transformations, although a number of other reactions also occur. Clays, organic matter, and metal oxides are capable of catalyzing abiotic reactions that occur in soils. Exchangeable cations can also influence the transformation of organic chemicals. Hydrolysis and redox reactions again dominate the abiotic reactions that occur in soils.

Microbial transformations of organic chemicals are classified as (1) biodegradation (contaminant used as substrate for growth, i.e., metabolism), (2) cometabolism (contaminant is transformed by metabolic reactions without being used as an energy source), (3) accumulation (contaminant is incorporated into the microorganism), (4) polymerization or conjugation (contaminant is bound to another organic chemical), and (5) secondary effects of microbial activity (contaminant is transformed due to indirect microbial effects, i.e., pH, redox) (Bollag and Liu, 1990). Although these transformations are considered to be mediated by microorganisms, abiotic transformations are also involved, especially in the transformations related to categories (4) and (5).

Biodegradation is considered the primary mechanism in which organic chemicals are transformed to inorganic products such as CO_2, H_2O, and mineral salts. The metabolism of organic chemicals in soils by bacteria is typically greater than for other microorganisms. In natural ecosystems, biodegradation of organic chemicals is facilitated primarily by heterotrophic bacteria and actinomycetes, some autotrophic bacteria, fungi including basidiomycetes and yeasts, and specific protozoa. Biodegradation can occur under aerobic and anaerobic conditions as shown for dichlorodiphenyltrichloroethane (DDT) in Figure 8-5.

8.4 STRATEGIES FOR REMEDIATING CONTAMINATED GROUNDWATERS AND SOILS

Contamination of our air, soils, streams, and groundwaters has become a national concern. Some of the more common contaminant problems are associated with leaking underground storage tanks, improper disposal of lubricants and solvents, industrial sludges, contaminant spills, as well as others. Sites which have been contaminated generally contain a mixture of organic constituents. It is therefore

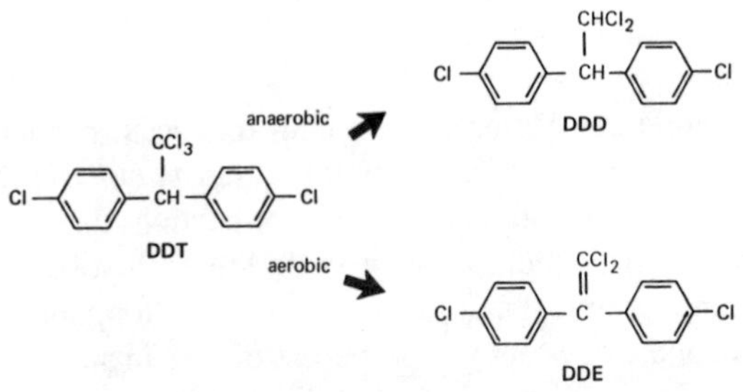

Figure 8-5   Example of anaerobic and aerobic products resulting from the biological transformation of the organic chemical, DDT.

important to understand how these multicomponent systems behave so that remediation strategies can be developed. In the previous sections, we discussed the various chemical, physical, and microbial processes that determine the movement and fate of organic chemicals. In the following sections, we will discuss several strategies used for remediating contaminated waters and soils.

The three stages of a remediation program are: identification, assessment, and action. Identification of a potential problem site requires that either the past history of the area and activities that took place are known, or an analysis of water and soil samples indicates a site has been contaminated. Assessment evaluates just how severe the contaminant problem is. In other words, is there a problem, where is the problem, and what is the extent of the problem? After an assessment is completed, a remediation action plan must be developed that will address the specific problems identified. A remediation action program may require that both soils, surface- and groundwaters be treated.

8.4.1 Groundwater Remediation Strategies

Several techniques are available for remediating contaminated groundwaters. If remedial action is considered necessary, three general options are available: containment, in-situ treatment, or pump-and-treat method (Figure 8-6). Methods used for the containment of contaminants will be discussed in Section 8.4.4, since many of these methods can be beneficial for restricting contaminant movement in either groundwater or soil environments. Of the remediation techniques, in-situ treatment measures are the most appealing because they generally do less surface damage; require a minimal amount of facilities; reduce the potential for human exposure to contaminants; are less expensive; and when effective, reduce or remove the contaminant.

The pump-and-treat method is one of the more commonly used processes for remediating contaminated groundwaters. With the pump-and-treat method, the contaminated waters are pumped to the surface where one of many treatment processes can be utilized. A major consideration in the pump-and-treat technology is the placement of wells; if the subsurface material has a low permeability, drains may be required instead of wells. Well placement is important and will depend on site characteristics (Table 8-10). Two types of wells, extraction and injection, are used. Extraction wells can be continuously pumping, pulsed pumping, and/or containment pumping. Injection wells are used in conjunction with extraction wells and are

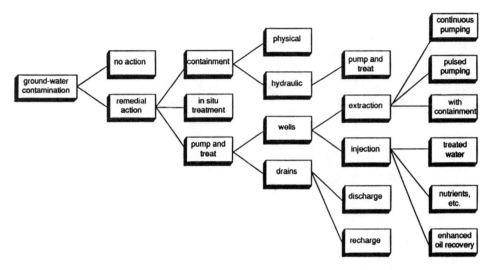

Figure 8-6   Options available for groundwater remediation, emphasizing the pump-and-treat methods. (Adapted from Mercer et al., 1990.)

positioned so that contaminant movement to extraction wells is enhanced. The injected solutions can be treated water, water containing nutrients and other substances that increase the chances for chemical or microbial degradation of the contaminants, or materials for enhanced oil recovery. An example of a site using injection and extraction wells is shown in Figure 8-7.

After the contaminated waters are pumped to the surface or collected in drainage systems, treatment is required to remove the contaminants. Treatment techniques can be grouped into three categories which include (1) *physical* — adsorption, separation, flotation, air and steam stripping, and thermal treatment; (2) *chemical* — ion exchange and oxidation/reduction; and (3) *biological* — land treatment, activated

Table 8-10   Important Hydrogeologic Characteristics of a Site That Are Used to Determine the Placement of Wells

Geologic
   Type of water-bearing unit or aquifer (overburden, bedrock)
   Thickness, areal extent of water-bearing units and aquifers
   Type of porosity (primary, such as intergranular pore space, or secondary, such as
      bedrock discontinuities, e.g., fracture or solution cavities)
   Presence or absence of impermeable units or confining layers
   Depths to water tables; thickness of vadose zone
Hydraulic
   Hydraulic properties of water-bearing unit or aquifer (hydraulic conductivity,
      transmissivity, storability, porosity, dispersivity)
   Pressure conditions (confined, unconfined, leaky confined)
   Groundwater flow directions (hydraulic gradients, both horizontal and vertical),
      volumes (specific discharge), rate (average linear velocity)
   Recharge and discharge areas
   Groundwater or surface water interactions; areas of groundwater discharge to surface water
   Seasonal variations of groundwater conditions
Groundwater use
   Existing or potential underground sources of drinking water
   Existing or near-site use of groundwater

Source: Mercer et al., 1990.

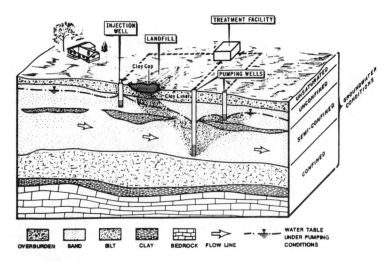

Figure 8-7   Site depicting the use of injection and extraction wells for remediation of a contaminated landfill site. (Adapted from Mercer, et al., 1990.)

sludge and aerated surface impoundments, and biodegradation. These treatment methods will be described in more detail in Section 8.4.3.

Due to the complex nature of subsurface environments, only limited success has been achieved with several of the in-situ methods; however, as our knowledge of these systems increases, we will undoubtedly hear of more in-situ remediation accomplishments. In-situ remediation of groundwaters can be achieved by chemical and biological techniques. Physical techniques such as adsorption retain the contaminant, which may be beneficial for further chemical and biological degradation. Biological in-situ techniques used for groundwater bioremediation can rely either on the indigenous (native) microorganisms, which may be stimulated with oxygen or nutrients, to degrade the contaminants or on amending the groundwater environment with microorganisms (bioaugmentation). Laboratory studies are currently being conducted with genetically engineered microorganisms capable of degrading a multitude of organic contaminants; however, before these microorganisms are used for in-situ groundwater remediation, additional research will be required to verify that they will not be harmful to the environment.

8.4.2 Remediation Strategies for Contaminated Soils

There are also several techniques for containing or treating contaminated soils. Figure 8-8 illustrates some of the options available if remediation is deemed necessary, although there are other techniques which may be more appropriate depending on the actual conditions of the contaminated site. With the excavation remedial action plan, soil and subsurface materials are dug up and either treated on-site or transported to another site. Several problems are associated with the latter method of cleanup: (1) transportation of the contaminant may present additional hazards such as loss of material, potential for accidents, and increased environmental disturbance; and (2) transportation costs may make the operation very expensive if the travel distance is

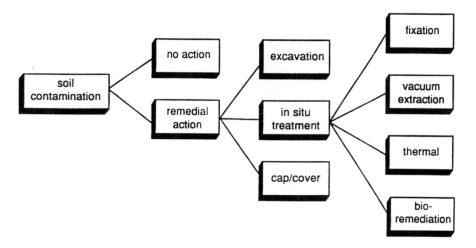

Figure 8-8   Options available for remediation of contaminated soils, emphasizing in-situ treatment methods. (Adapted from Mercer et al., 1990.)

great. Moving contaminated materials from one site to another does not solve the problem unless additional treatment steps will be taken. Land farming (see Chapter 9) may be a feasible option if the contaminant is degradable or useful for a practical purpose. However, incineration or chemical treatment of the contaminated material may result in problems that would also need to be corrected. Another concern related to transporting contaminated materials, would be how the public feels about accepting wastes; and in the case of incineration, what assurances does the public have that air pollution will not become a problem. The general feeling of many communities toward waste (nuclear, sludge, contaminated materials) disposal and incineration is "not in my backyard" (NIMBY).

If excavation and transportation is an unacceptable means of removing contaminated soils because of economical, social, or environmental reasons, other options (Figure 8-8) would include either containing the contaminated material or treating it on-site. The containment option may be the most feasible depending on the circumstances (i.e., nature of the contaminant, site conditions, hazard potential); containment methods used for restricting contaminant movement in soils and groundwater systems will be discussed in Section 8.4.4. Treating contaminated materials on-site also relies on excavation, but the remediation or treatment of the material is done without relocating the material to another site.

A typical on-site treatment procedure for processing contaminated materials requires first that the materials be excavated; second that the materials be treated by one of several methods to remove, destroy, or stabilize the contaminant; and third that the materials be replaced. Some of the more commonly used on-site treatment methods include extraction (leaching), separation, flotation, thermal treatment, steam stripping, chemical treatment, and microbial degradation. While all of these methods may be applicable for decontaminating soils containing organic chemicals, the most generally applicable methods are extraction, thermal treatment, and microbial degradation.

In-situ treatment methods for remediating organic chemical contaminated soils are gaining acceptance because they require less physical manipulation of the contaminated area and are generally less expensive. In-situ treatment is done without excavation and typically requires injection of a liquid, slurry, or gas that contains the chemical or microbial agent that will interact with the contaminant. Some of the in-situ methods that have been used or proposed are chemical treatment, solidification, thermal treatment, and microbial degradation. In-situ biodegradation involves injection of agents (e.g., oxygen H_2O_2, nutrients) to interact with the bacteria not the contaminant.

Another form of in-situ treatment of contaminated sites relies on the ability of plants to accumulate or detoxify organic chemicals. Although the use of plants for in-situ treatment has not been extensively studied, the fact that plants have the ability to absorb and metabolize herbicides would suggest there is probable merit to this type of in-situ treatment. Wetlands are currently being used for treatment of wastewater effluent, which contains varying amounts of organic material. Constructing artificial wetlands in or around a contaminated site may effectively assist in the decomposition of organic chemicals. Riperian buffer zones are a good example of how natural processes control the degradation of streams and rivers.

8.4.3 Treatment Methods

Treatment methods discussed in the previous sections are listed below. The various treatment methods are grouped according to the three basic techniques — physical, chemical, or biological.

8.4.3.1 Physical

Adsorption methods rely on the physical sorption or trapping of organic contaminants on activated charcoal or a synthetic-based resin. Contaminated waters are passed through columns or large vessels filled with the resins. Activated charcoal is preferentially used and is highly effective in removing low solubility organic chemicals.

Separation methods include treatments that physically separate contaminants from soils, and reserve osmosis, which concentrates contaminants by forcing water through a semipermeable membrane using pressure. Physical separation of aqueous and organic phase contaminants is achieved by the use of pressure or suction. Low-molecular-mass alcohols, ketones, amines, and aldehydes can be separated using the reverse osmosis method.

Flotation, or density separation, is commonly employed to separate low-density organic chemicals from soils and groundwaters. To facilitate the flotation process, chemical agents can be added to the soils or waters to release organic chemicals that adhere to solids. Flotation has also been used to recover oil from oil sands and tar sands.

Air and steam stripping can remove volatile organic chemicals from both soils and waters. The volatile organics discharged can be collected by trapping them on activated charcoal or by use of a condenser. The method is particularly useful for removing water soluble hydrocarbons (e.g., methanol, ethanol, isopropanol, phenol), water immiscible hydrocarbons (e.g., benzene, toluene, xylene), and halogenated hydrocarbons (e.g., trichloroethylene, methylene, dichlorobenzene).

Thermal treatment relies on heat to remove the contaminants either by evaporation or by destroying the contaminant with sufficient heat (incineration). In the former method, if significant amounts of highly volatile organic chemicals such as hydrocarbons and halogenated hydrocarbons are present, they possibly can be recovered. Complete incineration of organic chemicals in liquids, gases, and solids will produce CO_2, H_2O, HCl, and other products.

Solidification can be accomplished by physically or chemically binding the soil or waste material into a solid mass. Contaminated material can be solidified directly, or the materials around a contaminated site can be solidified to form a barrier. Solidification treatments often include the use of cement and plastic binding materials.

8.4.3.2. Chemical

Chemical treatment involves the addition of a chemical agent either on the soil surface or through an injection system that allows deeper penetration of the agent into the subsurface environment. The chemical agent can react with the organic contaminant to neutralize, immobilize, or chemically alter the contaminant so that it is harmless to the environment. In certain-situations, leaching a contaminated site with water or salt solution may provide enough protection from further degradation. However, this method is not applicable where the leachates can migrate into groundwaters.

Extraction (leaching) of organic chemicals such as hydrocarbons and halogenated hydrocarbons may be accomplished by using one of several different aqueous extracting agents such as an acid, base, detergent, or organic solvent miscible in water. Organic chemicals are known to be sorbed by, or partition into organic surfactants and the organic matter phase of soils. Some basic extracting agents such as NaOH or Na_2CO_3 are capable of dissolving or dispersing the organic matter, which will enhance the removal of organic chemicals. Organic solvents that have been proposed for extracting contaminated soils must be miscible with water to facilitate their separation after treatment. Suggested organic solvents include ethanol, isopropanol, and acetone.

Oxidation of organic contaminants has been extensively used in the treatment of groundwaters. Some of the more commonly used chemical oxidizers are air, oxygen, ozone, ozone plus ultraviolet (UV) light, chlorine, hypochlorite, and hydrogen peroxide. In situ oxidation treatment of soils and groundwaters has been accomplished by injecting oxygen-rich waters into the contaminated areas. Presumably, the enhanced degradation of the organic contaminants was by indigenous microorganisms.

Ionic and nonionic exchange resins can adsorb contaminants, reducing their leaching potential. Organic contaminants that possess charged sites, i.e., organic acids (carboxylic acids and phenols) and organic amines, as well as hydrophobic organics, are adsorbed by exchange resins. By injecting exchange resins into the subsurface zone, contaminants that are not effectively adsorbed by the subsurface materials can be retained possibly long enough for chemical or microbial degradation to render the organic chemical harmless.

8.4.3.3. Biological

Land treatment is an effective method for treating contaminated groundwaters and soils. The contaminated materials are mixed into or dispersed over the surface of the soil. Microbial degradation of the organic chemicals is enhanced over time as the microorganisms become acclimated to the contaminant. Methods for applying groundwaters to the land farming site include irrigation, overland flow, or subsurface irrigation. Additional discussion on land farming is presented in Chapter 9.

Activated sludge and *aerated surface impoundments* are used to degrade organic contaminants present in water. Both methods facilitate the degradation of organic contaminants by enhancing microbial degradation. In the activated sludge treatment process, the sludge material, which is rich in microorganisms capable of degrading organic chemicals, is recycled. Depending on the nature of the surface impoundment, both aerobic and anaerobic degradation processes can be operative.

Biodegradation, or microbial degradation, is one of the five biological mediated processes which transform organic contaminants, and in many cases is more important than physical and chemical processes. Biodegradation of contaminated soils and groundwaters is a relatively new technology which has gained considerable attention for in-situ remediation. Several studies, both in the United States and Europe, have shown biodegradation to be effective in remediating contaminated sites. However, biodegradation rates of some organic chemicals such as high-molecular-mass hydrocarbons, polynuclear aromatics, and highly substituted organic compounds (e.g., PCB) can be relatively slow. Additional discussion on the bioremediation of soils contaminated with organic chemicals is given in Chapter 9.

8.4.4 Methods Used for the Containment of Contaminants

Isolating contaminants by placing barriers above, below, or around them is a form of remediation that further restricts movement of the contaminant. Although containment systems usually possess a lower permeability than the soil and subsurface materials surrounding the contaminated site, they do not provide a 100% assurance that future movement of contaminants will not occur. Some of the types of containment barriers currently in use are the slurry trench wall, grouting curtain, vibrating beam wall, sheet piling, and bottom sealing. Containment barriers are often constructed of soil, clay, concrete, plastic, steel, or other grouting materials. In the past, landfills were not required to place contaminant barriers around the perimeter of the landfill site. In 1979, EPA guidelines for leachate control in landfills suggested the bottom of the landfill should be at least 1.5 m above the high level of the water table; placing a barrier on the bottom and sides to seal off the landfill (if landfill conditions were conducive for groundwater contamination); and if a liner was deemed necessary, it was to be constructed of materials that have a permeability of 10^{-7} cm/sec (approximately 0.1 ft/year) or less.

Recent studies have shown that certain organic chemicals, adsorbed onto soil and clay materials, can provide an effective sorption barrier to the leaching of organic contaminants. The organic chemicals used for these studies were organic cations from a group of organic chemicals (quaternary ammonium cations, also known as

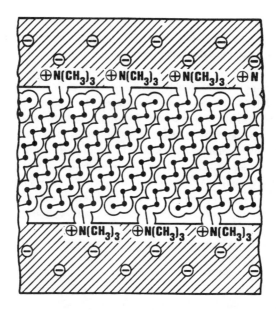

Figure 8-9 Example of an organo-clay. The organic cation $[(CH_3)_3N(CH_2)_{15}CH_3]^+$ is adsorbed in the interlayer region of a swelling clay. Nonionic organic contaminants such as benzene, toluene, and xylene are effectively adsorbed by the organo-clay materials. (Adapted from Boyd et al., 1991. With permission.)

QUATS) that are generally inexpensive and widely used in such products as detergents, fabric softeners, antistatic sprays, and swimming pool additives. Being positively charged, these chemical modifiers bind strongly to negatively charged sites in soils and form effective sorptive phases for organic contaminants. Modified soils and subsurface materials can be formed in-situ by injecting QUATS into an area around contaminated sites, or downgradient from a contaminant plane in order to prevent contaminant migration. In addition, organo-clays can be prepared and injected as a slurry to form an effective barrier around sites as well. Since organo-clays are capable of sorbing petroleum constituents (such as benzene, toluene, ethylbenzene, p-xylene, butylbenzene, and naphthalene), they would also be useful as a liner material around petroleum tank farms and underground storage tanks and possibly for the treatment of petroleum-contaminated waters. An example of an organo-clay is shown in Figure 8-9.

8.5 PEST MANAGEMENT

A pest control program must rely on an integration of pest management techniques and practices that assist in providing economic, ecological, and sociological benefits. The definition of "integration of pest management techniques and practices" refers to the compatible use of *biological, chemical, mechanical,* and *cultural controls* to manage pest populations. Pest management programs, as well as conservation tillage practices, are becoming more accepted within the agricultural production sector as individuals gain a greater understanding and appreciation of how nature works and as the concern for environmental quality grows.

As we move toward conserving energy and protecting our resources, we have implemented farming practices that can be both beneficial and detrimental. For example, conservation tillage systems have become popular farming methods because they generally result in a reduction in soil loss due to wind and water erosion and an increase in the amount of water retention when crop residues are left on the soil surface. However, some conservation tillage practices actually enhance problems associated with weeds, diseases, insects, nematodes, rodents, and soil microorganisms. Alternative methods such as biological control measures, use of resistant varieties, or altered cultural practices may be required in addition to applications of pesticides to combat the problems developed by conservation tillage systems.

8.5.1 Integrated Pest Management

Integrated pest management, also known as IPM, is a program of pest control that relies on several practices to prevent pest outbreaks from occurring. Preferably, these practices should be compatible and augment the effectiveness of each other. Some of the components of an effective IPM program are soil preparation for the control of pests (i.e., weeds, microorganisms, and insects), chemical (pesticides) or biological pretreatment for the control of weeds and insects, observations of insect activities both locally and regionally to better time management strategies, and understanding climatic conditions that are conducive to pest outbreaks.

In an IPM program, several practices must be followed to prevent pest problems. Practices commonly implemented in an IPM program include: use of certified disease- and insect-free seeds or plants; implementation of cultural practices such as crop rotation and sanitation measures; control of physical conditions (time of tillage, planting conditions, and temperature and moisture of storage conditions for the prevention of diseases); utilization of chemical practices (pesticides, fumigants, seed and plant treatments, and use of disinfectants); and development of innovative biological control practices, such as use of insect and disease resistant varieties (see the next section for additional discussion on biological control measures). Individually these practices may not provide adequate protection for pest control; however, collectively these practices can minimize economic losses due to pest problems.

8.5.2 Alternative Pest Control Measures

Alternatives to the use of pesticides for controlling weeds and insects have gained popularity in recent times because they tend to require less input and are more sustainable. Biological control is one alternative to pesticides, although biological control programs are usually designed to be host specific and therefore are not capable of solving multiple pest problems. In an ideal biological control program, pest populations are kept below levels where economic losses occur, without placing undue stress on the ecosystem. Hoy (1989) summarized the three commonly employed tactics used for biological control strategies as follows:

- *classical* (importation and establishment of exotic natural enemies to control exotic and occasionally native pests)
- *conservation* (actions to protect, maintain, and/or increase the effectiveness of natural enemies)
- *augmentation* (actions taken to increase populations or beneficial effects of natural enemies, which may not be self-sustaining)

Humans are primarily responsible for the dissemination of weeds and insects throughout the world. Many of the most troublesome weeds and insects in the U.S. (i.e., Russian thistle, St. John's wort (Klamath weed), and Johnsongrass; and Gypsy moth, screwworm, and Russian wheat aphid, to name a few) are a result of travel and commerce activities. Weeds and insects from exotic countries are capable of proliferating if: (1) they adapt to the new environment and (2) their natural enemies are not present. One approach for controlling weed and insect populations is to introduce biological controls from the pest's native region that can reduce or regulate the pest to a level that is economically and aesthetically acceptable. Natural enemies have been introduced to combat exotic weeds and insects that have displaced the native plant or insect species.

Biological control has been successfully used in the control of terrestrial and aquatic weeds. The biological control of the common prickly pear *(Opuntia inermis)* and spiny prickly pear *(O. stricta)* in Australia and of St. John's wort or Klamath weed *(Hypericum perforatum)* in the U.S. are classic examples of terrestrial weed control by beneficial insects that feed on these weed species and not on agronomic plants. Herbivorous (e.g., white amur [*Ctenopharyngodon idella*]) and nonherbivorous (e.g., carp [*Cyprinus carpio*]) fish have been found to control aquatic weeds either by consuming or uprooting them, respectively. Several examples of weeds common to the northwestern U.S. and biological control agents showing promise in their control are listed in Table 8-11.

Biological control of insects has also had some success. A significant example of biological insect control is the case of the vedalia beetle *(Rodolia cardinalis)* control of the cottony-cushion scale *(Icerya purchasi)* which threatened the livelihood of the California citrus industry in the 1880s. Other pests that were at least partially controlled by beneficial insects include the European corn borer *(Ostrinia nubilalis)*, European spruce sawfly *(Diprion hercyniae)*, and Oriental fruit fly *(Dacus dorsalis)*, to name a few. Successful biological control of pests can be accomplished on a small scale as well, such as in gardens or on small farms. Some of the more common examples of biological control in small-scale agriculture in the western U.S. are listed in Table 8-12.

Table 8-11   Some Biological Control Agents That Are Showing Promise for Weed Control in the Northwestern U.S.

Weed	Bioagent	Feeding habit
Canada thistle	*Orellia ruficauda*	Larvae consume seeds in developing seed heads
Diffuse and spotted knapweed	*Urophora affinis* *U. quadrifasciata*	Galls created in seed head by feeding of larvae
Gorse	*Exapion ulicis* *Agonopterix nervosa*	Weevil larvae eat seeds, adults eat foliage Larvae feed inside of growing tips of shoots
Italian, milk, musk, and slenderflower thistle	*Rhinocyllus conicus*	Weevil larvae consume developing seeds in head
Mediterranean thistle	*Phrydiuchus tau*	Larvae hollow out crown and burrow into root
Poison hemlock	*Agronoterix alstroemeriana*	Larvae defoliate leaves, stress plant
Rush skeletonweed	*Cystiphora schmidti* *Eriophyes chondrillae* *Puccinia chondrillina*	Midge larvae feeding causes leaves to bunch Mite feeding causes buds to gall, rust forms sori on underside of leaves; plant disease
St. John's wort	*Chrysolina quadrigemina* *C. hyperici* *Agrilus hyperici*	Beetle defoliate plants reducing photosynthesis Beetle larvae tunnel inside root
Scotch broom	*Apion fuscirostre*	Weevil larvae eat seeds; adults eat foliage
Tansy ragwort	*Tryia jacobaeae* *Pegohylemyia seneciella* *Longitarsus jacobaeae*	Moth larvae feed on leaves defoliating plant Larvae consume seeds in developing seed heads Flea beetle larvae feed on root bark and mine crown
Yellow star thistle	*Bangasternus orientalis*	Weevil larvae consume developing seeds in head
Yellow toadflax	*Calophasia lunula* *Gymnaetron antirrhini* *Brachypterolus pulicarius*	Larvae consume foliage Galls created in seed head by feeding of larvae; adults feed on stems; beetle larvae feed on petals and pollen

Source: Burill et al., 1992.

Table 8-12 Biological Control of Some Common Insect Pests of the Western U.S.

Bioagent	Target	Comments
Predators		
Lady beetles (Coccinellidae)	Aphids and spider mites	Voracious; consume soft-bodied insect/eggs
Ground beetles (Carabidae)	Insects	Prey on most insects found on the soil surface
Rove beetles (Staphylinidae)	Small insects	Control seed corn and onion maggots
Green lacewing (Chrysopidae)	Caterpillars and beetles	Feed on aphids, and insects that are sometimes larger than the lacewing
Syrphid fly (Syrphidae)	Aphids	Important aphid control during cool part of the growing season; larvae are predatory stage
Predatory bugs (Hemiptera)	Insects and mites	Pierce prey and suck out body fluids
Stink bugs (Pentatomidae)	Potato beetle larvae	Feed on other insects as well
Assassin bugs (Reduviidae)	Caterpillars and beetles	Not very abundant
Damsel bugs (Nadidae)	Aphids	Adults and nymphs feed on aphid eggs and larvae
Minute pirate bugs (Anthocoridae)	Thrips and spider mites	Feed on wide variety of insect eggs
Hunting wasps (Sphecidae)	Caterpillars and beetles	Selective for certain prey; females paralyze prey for larvae to feed on
Spiders (Araneida)	Living insects and small anthropods	General insect predators; effective bioagents
Parasites		
Tachinid flies (Tachinidae)	Caterpillars, beetles, andother bugs	Young fly maggots tunnel into host and feed on them; most are restricted to particular hosts
Parasitic wasps Braconid wasps (Braconidae) and Icheumanid wasps (Ichneumonidae)	Aphids and small insects	Develop inside host from eggs inserted by female wasp; host dies when parasite emerges
Chalcid wasps (Chalcidoidea)	Aphids	Attack caterpillars of cutworms, fall webworms, and cabbage loopers; many are egg parasites
Insect diseases		
Viruses	Caterpillars and sawflies	Viral diseases are widespread in nature; often used in spray formulations to control pests such as spruce bud worm
Bacteria	Caterpillars, moths, hornworms, leaf rollers, cutworms, fruit worms, European corn borer	*Bacillus thuringiensis* the most notable and sold commercially for lepidopteran control in gardens and agricultural crops, also effective against mosquito larvae
Fungi	Insect and mites	Attempts to develop fungi that can control mosquitoes and greenhouse pests are currently under way

Source: Cranshaw, 1992.

REFERENCES

Bollag, J. M. and Liu, S. Y., Biological transformation processes of pesticides, in *Pesticides in the Soil Environment: Processes, Impacts, and Modeling,* Cheng, H. H., Ed., Soil Science Society of America, Madison, WI, 1990, 103.

Boyd, S. A., Jaynes, W. A., and Ross, B. S., Immobilization of organic contaminants by organo-clays: application to soil restoration and hazardous waste containment, in *Organic Substances and Sediments in Water,* Baker, R. S., Ed., Lewis Publishers, Chelsea, MI, 1991, 181.

Burill, L. C., William, R. D., Parker, R., Boerboom, C., Callihan, R. H., Eberlein, C., and Morishita, D. W., *Pacific Northwest Weed Control Handbook,* Agricultural Communications, Oregon State University, 1992.

Chiou, C. T., Freed, V. H., Schmedding, D. W., and Kohnert, R. L., *Environ. Sci. Technol.,* 11, 475, 1977.

Cranshaw, W., *Pests of the West,* Fulcrum Publishing, Golden, CO, 1992.

Cullimore, D. R., Interaction between herbicides and soil microorganisms, *Residue Rev.,* 35, 65, 1971.

Harris, J. C., Rate of hydrolysis, in *Handbook of Chemical Property Estimation Methods: Environmental Behavior of Organic Compounds,* Lyman, W. J., Reehl, W. F., and Rosenblatt, D. H., Eds., McGraw-Hill, New York, 1982a, chap. 7, 48.

Harris, J. C., Rate of aqueous photolysis, in *Handbook of Chemical Property Estimation Methods: Environmental Behavior of Organic Compounds,* Lyman, W. J., Reehl, W. F., and Rosenblatt, D. H., Eds., McGraw-Hill, New York, 1982b, chap. 8, 43.

Hoy, M. A., Integrating biological control into agricultural IPM systems: reordering priorities, Proc. Natl. Integrated Pest Manage. Symp./Workshop, Las Vegas, NV, Commun. Serv. NY, State Agricultural Experimental Station, Cornell University, Geneva, NY, 1989, 41.

Mabey, W. and Mill, T., *J. Phys. Chem.,* 7, 383, 1978.

Madhun, Y. A. and Freed, V. H., Impact of pesticides on the environment, in *Pesticides in the Soil Environment: Processes, Impacts, and Modeling,* SSSA Book Series No. 2, Cheng, H. H., Ed., Soil Science Society of America, Madison, WI, 1990, 429.

Marathon-Agricultural and Environmental Consulting, *Video cassettes — Fate of Pesticides in the Environment,* Box 6969, Las Cruces, NM 88006, 1992.

Mercer, J. W., Skipp, D. C., and Giffin, D., Basics of Pump-and-Treat Ground-Water Remediation Technology, U.S. Environmental Protection Agency, EPA/600/8-90/003, Ada, OK, 1990.

Miller, D. W., Ed., *Waste Disposal Effects on Ground Water,* Premier Press, Berkeley, CA, 1980, 512.

Office of Technology Assessment, Protecting the Nation's Groundwater from Contamination, U.S. Congress, Office of Technology Assessment, OTA-0-233, U.S. Printing Office, Washington D.C., 1984.

Parsons, D. W. and Witt, J. M., Pesticides in the Groundwater in the United States of America: A Report of a 1988 Survey of State Lead Agencies, Oregon State University Extension Service, Corvallis, OR, 1988.

Shoemaker, C. A., Integration of environmental concerns into IPM programs, Proc. Natl. IPM Symp./Workshop, Las Vegas, NV, Commun. Serv. NY, State Agricultural Experimental Station, Cornell University, Geneva, NY, 1989, 121.

Thomas, R. G., Volatilization from water, in *Handbook of Chemical Property Estimation Methods: Environmental Behavior of Organic Compounds,* Lyman, W. J., Reehl, W. F., and Rosenblatt, D. H., Eds., McGraw-Hill, NY, 1982, chapter 15, 34.

Weber, J. B. and Miller, C. T., Organic chemical movement over and through soil, in *Reactions and Movement of Organic Chemicals in Soils,* Sawhney, B. L. and Brown, K., Eds., Soil Science Society of America, Madison, WI, 1989, 305.

SUPPLEMENTARY READING

Cheng, H. H., Ed., *Pesticides in the Soil Environment: Processes, Impacts, and Modeling,* Soil Science Society of America, Madison, WI, 1990.

Barcelona, M., Wehrmann, A., Keely, J. F., and Pettyjohn, W. A., Eds., *Contamination of Ground Water: Prevention, Assessment, Restoration,* Noyes Data Corp., Park Ridge, NJ, 1990, 213 pp.

Calvet, R., Adsorption of organic chemicals in soils, *Environ. Health Perspect.,* 83, 145, 1989.

Lyman, W. J., Reehl, W. F., and Rosenblatt, D. H., Eds., *Handbook of Chemical Property Estimation Methods: Environmental Behavior of Organic Compounds,* McGraw-Hill, New York, 1982.

Smith, M. A., Ed., *Contaminated Land: Reclamation and Treatment,* Plenum Press, NY, 1985, 433 pp.

Thomas, R. G., Volatilization from soil, in *Handbook of Chemical Property Estimation Methods: Environmental Behavior of Organic Compounds,* Lyman, W. J., Reehl, W. F., and Rosenblatt, D. H., Eds., McGraw-Hill, New York, 1982, chap. 16, 50.

Wolfe, N. L., Mingelgrin, U., and Miller, G. C., Abiotic transformations in water, sediments, and soil, in *Pesticides in the Soil Environment: Processes, Impacts, and Modeling,* Cheng, H. H., Ed., Soil Science Society of America, Madison, WI, 1990, 103.

9   BIOGEOCHEMICAL CYCLES AND SOIL MANAGEMENT

9.1 BIOGEOCHEMICAL CYCLES AND THE ENVIRONMENT

All of the nutrients and trace elements that we have considered in earlier chapters (e.g., N, P, S, and trace elements), as well as the pesticides and airborne pollutants discussed in Chapters 8 and 10, can undergo physical, chemical, and biological transformations in soils. These transformations can increase, reduce, or even have no effect on the environmental impact of each of these elements or compounds. A *biogeochemical cycle* can be defined as a conceptual description of the mechanisms by which an element or compound is transformed within a system of interest including the means by which the various forms are interchanged between the solid, liquid, and gaseous phases of that system. Biogeochemical cycling is of course not restricted to the soil environment, but includes geologic materials, biological organisms, air, and waters, which are fundamentally global in nature; however we often view and attempt to manage them at smaller scales such as within a watershed or even within a city or a farm. These cycles not only summarize the major processes involved, but provide a generalized overview of the environmental factors that control each transformation. The global and soil N cycles, illustrated earlier in Figures 4-1 and 4-2, are good examples of the different scales of biogeochemical cycling of an important element for soil management. We understand the general types of processes operative in the global N cycle and many of their environmental impacts (e.g., groundwater contamination, ozone depletion, acid rain); we seek to manipulate the soil N cycle to minimize these negative effects.

Specifically, from the perspective of soil science, we wish to use our knowledge of biogeochemical cycles to monitor and control the environmental fate and transport of soil constituents. Earlier chapters have described many of these chemical and biological transformations (e.g., adsorption-desorption, precipitation, mineralization-immobilization, oxidation-reduction) and the factors that affect the transport of an element or compound between different phases of the cycle via leaching, erosion, runoff, or volatilization. Consequently, biogeochemical cycles often become the basis for modeling efforts that seek to identify the most likely fate(s) of plant nutrients, nonessential elements, or organic molecules. By improving our understanding of biogeochemical cycles, we hope to identify management practices that can minimize potential environmental degradation. We rely on scientific research to define the possible and most likely transformations that can occur in soils. From this research we seek management practices that provide us some measure of control over these transformations in order to maximize their benefits and minimize their risk. One of the critical steps in this process is the realization that for most elements and compounds, the types of transformations possible and their rates are limited by the range in properties of the system in which they are located. As a result, we do not have to manage (and often cannot) every possible biochemical reaction and transport process; instead we must prioritize our efforts. There are many examples of this. Transformations that only occur under anaerobic conditions (e.g., denitrification) are common in a wetland, but highly unlikely in arid zone soils; leaching of P is rare in most agricultural soils, but may become a problem in sandy soils used for intensive wastewater irrigation; trace element deficiencies are less common in humid regions because the extensive leaching of basic cations results in acidic, not alkaline soils and greater solubility of most metals; and biodegradation of organic pollutants may occur rapidly in surface soils, but slowly in subsoils where many of the nutrients required by soil microorganisms involved in the degradation process are found in very low concentrations. Understanding (and quantifying to the extent possible) the constraints placed on the biogeochemical cycling of an element in different systems is an important first step in the development of sound environmental management practices for soils.

Frequently, however, despite our concerns about the many possible impacts a potential pollutant may have on the environment, we lack the scientific knowledge, the technology, or the resources to control every possible transformation within a biogeochemical cycle. Prioritization of management efforts requires that we combine our understanding of these cycles with an assessment of the environmental "risk" associated with each transformation (see Chapter 11 for complete discussion of the risk assessment process). This chapter will discuss some differences between biogeochemical cycles in natural, agricultural, urban, and disturbed soils; and will illustrate how we currently use our knowledge of these cycles and the risks associated with them to manage fertilizers, organic wastes, and biodegradation of organic pollutants.

9.1.1 Biogeochemical Cycles: Bioregions, Biomes, and Ecosystems

Perhaps the broadest perspective that can be taken when considering the importance of biogeochemical cycles is that of the *bioregion*, a geographic area where land use is defined by the natural resources present and is thus limited by the soils, geology, topography, and climate characteristic to the region. A related concept is the *biome*, an ecological region with similar types of biological organisms and physical environments. The major biomes of the world are shown in Figure 9-1. Based on the properties of each biome (e.g., tundra, temperate forest, desert) we can anticipate characteristic patterns in biogeochemical cycling of elements and use research and management to identify generally appropriate, or inappropriate, land uses. On a smaller scale, the interrelationships and transfers between the distinct *ecosystems* within a biome must also be considered if we are to develop land management

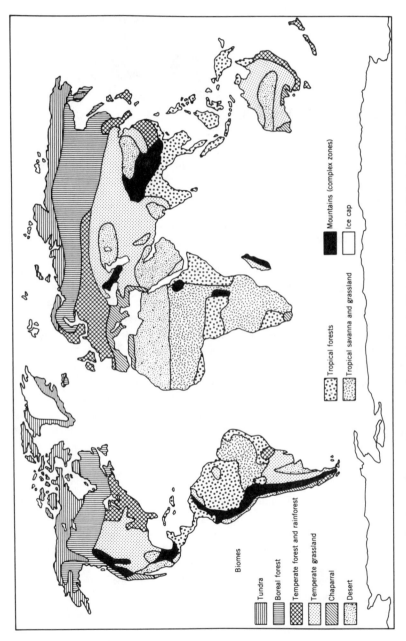

Figure 9-1 Geographic perspective on the major biomes of the world. (Adapted from Council on Environmental Quality, 1989.)

Biomes

Tundra

Boreal forest

Temperate forest and rainforest

Temperate grassland

Chaparral

Desert

Tropical forests

Tropical savanna and grassland

Mountains (complex zones)

Ice cap

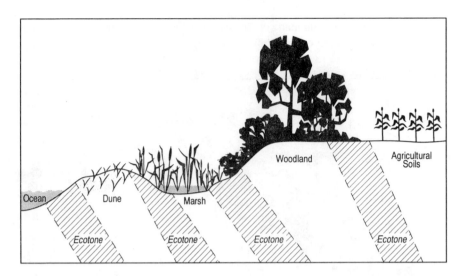

Figure 9-2   Interacting ecosystems in the landscape, showing the ecotones (transitional zones) that exist between individual ecosystems.

practices with minimal environmental impact (Figure 9-2). The term *ecosystem* means many things to many people; broad definitions are often similar to those used for bioregions, but on a smaller and more manageable scale; they refer to geographic areas with stable and reasonably similar biological communities, including the nonliving materials such as soil, air, and water. For our purposes an ecosystem refers to a region with sufficiently distinct biological and physical environments to be adaptable to similar approaches of land-use management. It is beyond the scope of this book to describe all possible biomes and ecosystems and the transfers of matter and energy that occur between these regions. We will focus primarily on the relationship between biogeochemical cycling in "managed soils," those used for agriculture or waste disposal, and their possible effects on nearby "natural ecosystems" such as wetlands.

9.1.2 Biogeochemical Cycles: Interacting Ecosystems and Ecotones

Many natural ecosystems such as upland forests, wetlands, and deserts are located in close proximity to soils used for agricultural production, cities and industry, mining, transportation systems, and waste disposal sites. Understanding how adjacent ecosystems impact one another requires an understanding of biogeochemical cycling on a more regional scale. A broad example of interacting ecosystems would be the agricultural production areas and natural wetlands located near urban areas. Agriculture provides the food and fiber for the municipality and is often relied on to accept and manage many of the waste materials generated by the population and the industries in the urban ecosystem. The urban areas provide the market for agricultural products; the industrial and financial infrastructures that sustain agribusiness; and other "ecosystem outputs," such as education, scientific research, recreational activities, and goods and services. Wetlands perform many important functions for both of these ecosystems. Among many other functions, they act as natural filtration zones for agricultural and urban runoff; they are important as a means to accept and retain water during storm events, thus preventing flooding; and they maintain diverse and unique habitats for wildlife and plants.

Increasingly, however, we find that the stability of natural ecosystems can be affected negatively by anthropogenic inputs, a ready example being the concern for the effect of acidic deposition from the burning of fossil fuels on the ecology of forests and lakes. Similarly, in some cases we now use natural ecosystems to mitigate the environmental impact of other activities, such as when we rely on wetlands to purify polluted waters from agricultural runoff, municipal wastewater treatment plants or surface and underground mine drainage. The sustainability of the wetland ecosystem when it interacts with the urban ecosystem then becomes an environmental issue. In each of these examples the nature of the biogeochemical cycling of elements (such as N, P, and S) and or organic compounds (such as pesticides and industrial organic wastes) must be defined in distinctly different ecosystems if we are to develop management strategies that sustain both.

In summary, to maintain both natural and human-dominated ecosystems we must understand what the nature of the present internal biogeochemical cycles is and how human activities can alter the cycling and perhaps the stability of each system. We also need to be aware of the boundaries that exist between ecosystems, natural and human-made, that inhibit or facilitate transfers between the geographic areas we are capable of managing. The complex nature of most ecosystems means that these boundaries are often not distinct. We use the term *ecotone* to denote the transitional zones present between ecosystems. A good example of an ecotone is a wetland, an ecosystem that has some properties of the zones bordering it such as the upland forest, and the dune system (Figure 9-2). Another example of an ecotone might be *greenspace,* the parks, woodlots, and wetlands interspersed throughout a densely populated urban area that buffer many of the negative effects of an urban environment (such as noise, traffic, odors, heat, and storm water runoff).

The remainder of this chapter will focus on examples of how we use our understanding of the biogeochemical cycling of nutrients and trace elements to develop environmentally sound management programs that sustain soil productivity and ecosystem stability. A brief section is also included on bioremediation to illustrate a somewhat different approach to management of a biogeochemical cycle that is used when an ecosystem is damaged by large-scale contamination.

9.2 MANAGEMENT OF BIOGEOCHEMICAL CYCLES

The sheer number of individual biogeochemical cycles present in the soil alone is staggering. There are 103 elements in the periodic table, each with a distinct cycle. Some elements (C, H, O, N, P, S) cycle rapidly and ubiquitously because they are components of biological organisms (*biogenic* elements); others such as the trace elements are cycled at slower rates and in more localized geographic areas. Tens of thousands of synthetic organic compounds are produced in significant quantities by industries each year; many enter the soil environment as products (e.g., pesticides), as constituents of industrial and municipal wastes and wastewaters, or as accidental spills and leaks during transportation and storage. Each individual cycle is complex and often closely related to transformations undergone by other elements or compounds present. Nitrogen cycling, for example, can affect biodegradation of soils contaminated with organic chemicals because N is an essential nutrient for the microorganisms that degrade the chemical. Carbon cycling affects the availability of N, P, and S to plants and the mobility of trace elements in soils. Complicating the matter further is the fact that the bioregion or individual ecosystem in which the cycles are located determines which processes within a cycle predominate in the soil. Indeed, one of the more challenging aspects of soil management, for either agricultural or environmental purposes, is the highly site-specific nature of many management programs. Examples of this are plentiful: we cannot manage fertilizer N in exactly the same way in all soils because of the interactions between soil type and N losses; a reclamation strategy that successfully revegetates a mine spoil in the southeastern United States may fail completely in the mountains of Colorado because of climatic differences; enhancing the biological degradation of gasoline from leaking underground storage tanks may be possible on-site in one state, while in another it may require excavation and incineration. If we have learned anything about soil management, it is the fact that managing biogeochemical cycling requires not only a good scientific understanding of the processes operative in the cycle, but also ingenuity and innovation. Indeed, it could be reasonably argued that we do not manage these cycles at all; instead we simply manipulate a limited number of factors to redirect one or more of the processes to our advantage. Perhaps this is why many individuals find environmental soil science an interesting scientific discipline — it provides an opportunity to adapt basic science (soil chemistry, physics, and microbiology) and practical knowledge to important and difficult problems faced by our society.

The following sections will describe some current approaches used to manage biogeochemical cycles, under "real-world" conditions. We will consider nutrient management for farms with and without animals; municipal waste use for agricultural crop production; nutrient transformations in wetlands; and bioremediation of an organic chemical accidentally discharged into the soil. The approaches used are generalized and often vary considerably from one ecosystem to the next. Most represent the cooperative efforts of multidisciplinary teams, a characteristic feature of all good soil management programs.

9.2.1 Nutrient Management for Farms with and without Animal Production

Nutrient management on a farm begins with a quantitative assessment of the current balance between nutrient inputs to the farm and nutrient outputs from the farm. In essence, we seek to determine whether farming practices are "overloading" some transformation in biogeochemical cycle resulting in a negative effect on an adjacent ecosystem. As shown earlier in Figure 5-18, nutrient inputs on farms include fertilizers, organic wastes (e.g., animal manures or wastewaters, municipal sludges or composts), feeds, and natural contributions from soils, irrigation waters, and atmospheric precipitation. Nutrients leave the farm in harvested grain, forage, or produce; in animal products; and by processes such as leaching, runoff, denitrification, and volatilization. Practically speaking, in most agricultural situations only a limited number of biogeochemical cycles are actually "managed" by farmers or those involved in advising farmers. Agricultural research has shown that the most important and readily manipulated nutrient cycles from a crop production standpoint are N, P, and K. The C cycle must also be considered because of its role in controlling the mineralization or immobilization of N and P; farmers, however, rarely actively manage the cycling of C in soils. Proper management of soil pH through liming usually ensures that the Ca and Mg cycles provide adequate amounts of these nutrients for plant growth without toxic effects from acidic cations (H, Al, Mn); S and trace element cycles are normally of less concern for crop management except in certain well-understood, localized conditions. As an example, S deficiency and significant crop response to applications of S fertilizers are most common in humid regions on deep, sandy soils; trace element deficiencies are usually restricted to alkaline or overlimed soils (Fe, Mn, Zn) or high organic matter soils (Cu).

A *nutrient budget* is a quantitative form of a biogeochemical cycle, one that estimates whether the current management practices on the farm are resulting in a nutrient deficit or excess. Nutrient budgets are normally constructed by first estimating the crop nutrient requirements at a realistic yield. Next the contributions of soil nutrients are assessed by a comprehensive soil testing program, one that may even include subsoil testing for some nutrients that may be found below the "plow layer", but within the rooting zone of the crop. A good example of a nutrient where subsoil testing is important is S. Many studies have shown that the sulfate form of S (SO_4^{2-}) only leaches to moderate depths in soils and may be an important reservoir of plant available S once roots penetrate to depths of 30–60 cm. Results from soil tests provide not only an estimate of the amount of a nutrient present in the soil in a plant available form, but also the likelihood of an economic crop response to fertilization with that nutrient and the rate of the nutrient needed to obtain optimum growth for the specified crop. The difference between crop nutrient requirements and that available from the soil represents the amount that must be provided from external sources. At this point any other significant sources of nutrient inputs, such as those provided in irrigation waters, should also be taken into account when developing the nutrient budget. For farms without a significant animal production component, the final step in the nutrient budget process is to determine the most economical source of nutrients available and the most efficient application technique. However, if a farm has a large animal-producing facility (such as a poultry operation or a cattle feedlot), the next step is to account for the amount of nutrients available "on-farm" from animal wastes, since these materials are almost always applied to cropland. Methods to estimate the amount of N available from these wastes were described in Chapter 4. Although less information is available on the rate of release of other elements from animal manures, most states provide estimates of the nutrient content of animal manures so that farmers can adjust fertilizer applications by proper crediting of manure nutrients, as shown in Table 9-1 for poultry manures. Detailed information on the nutrient content of animal manures is available from many sources; one example is the publication *Livestock Waste Management* (1983).

Table 9-1   Example of Nutrient Credits Used for Poultry Manures

Condition of poultry manure	Solids (%)	N	P$_2$O$_5$	K$_2$O	Ca	Mg (kg/Mg)	S	Mn	Cu	Zn
Liquid	5	5	4	2	3	T[a]	T	T	T	T
Moist and crumbly	50	20	20	10	35	3	2	0.2	0.2	0.2
Dry	85	45	35	20	70	10	4	0.5	0.5	0.5

Source: Pennsylvania State University, 1983.
[a] T = Trace quantity.

In many instances, the long-term use of animal wastes produces an excessive amount of certain nutrients in soils, particularly those soils in close proximity to the site of waste generation (e.g., P, see Table 5-6). If this nutrient, or other essential or nonessential element, can have a significant environmental impact, the nutrient management plan must include strategies to minimize that impact. Conversely, if a nutrient deficit exists, the most efficient use of on-farm resources and fertilizers to optimize the supply of all nutrients to the crop must be identified. The resolution of problems of nutrient excess that are identified by a properly developed, farm-wide nutrient budget can be challenging because of the lack of economic alternatives to manure use other than land application near the site of manure production. While this situation is most common with animal-based agriculture, it is by no means confined to that scenario. Other examples include farms that have high soil test P levels from long-term overfertilization with commercial fertilizers and farms operated by municipalities or industries for the purposes of waste disposal.

Implementation of the nutrient management plan based on the nutrient budget requires several steps including efficient storage, handling, and application of the nutrients to the most appropriate sites on the farm. The *site plan* shown in Figure 9-3 was developed for a dairy farm and illustrates many of the factors that must be considered to ensure that one ecosystem (the farm) does not adversely impact an adjacent ecosystem (a riparian zone and stream). It represents an integrated approach to managing the major biogeochemical cycles of importance to this farm (nutrients) through development of a comprehensive plan to store, handle, and distribute nutrients from manures and fertilizers to a variety of crops. Environmental protection is considered by the use of storm water runoff ponds, riparian corridors, and proper placement of animal and manure production facilities relative to drinking water wells.

9.2.2 Management of Municipal Wastes for Soils

Urban areas produce large quantities of a wide variety of organic waste materials that are believed to be suitable for land application programs. Wastes such as sewage sludge, composts of sludges and wood by-products, municipal solid waste composts (without sludge), yard waste (leaves, grass clippings) composts, and wastewaters from sewage treatment or industrial sources are commonly applied to cropland, forests, turf, and ornamentals grown in landscapes and roadsides and in large-scale land reclamation projects. Application of these materials to soils is almost always regulated by federal and state environmental agencies. Normally permits must be

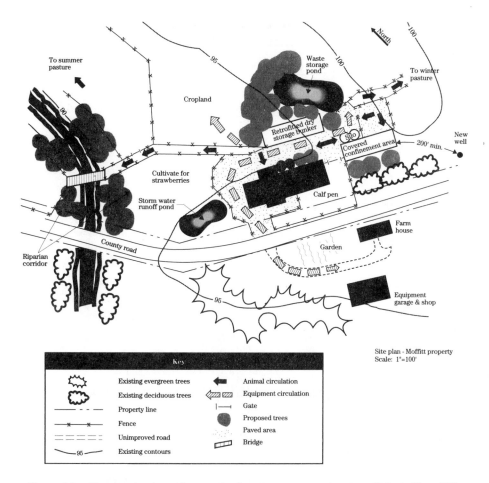

Figure 9-3 Site plan developed for an animal waste management system. (Adapted from U.S. Department of Agriculture, Soil Conservation Service, 1992.)

obtained from these agencies following a review process that requires detailed, site-specific information on all aspects of the land application program including — but not limited to — soil type, crop rotation, surface- and groundwater properties, topography, odor control, and monitoring programs. Unlike agricultural situations where only a few biogeochemical cycles must be managed and monitored, environmental regulatory agencies require that management plans for municipal and industrial wastes carefully consider a large number of inorganic and organic constituents in wastes, as illustrated in Table 9-2 for Delaware. Management and monitoring programs must, therefore, often be based on our understanding of the biogeochemical cycles of dozens of elements or compounds. Most regulatory agencies use the concept of *land-limiting constituents* (LLC) in wastes to determine both annual application rates and the total site life, or length of time a waste material can be applied to a site. The land-limiting constituent is the element or compound in a waste that is perceived to present the greatest hazard and that should be used to determine

Table 9-2 Monitoring Requirements for Sewage Sludges
and Sludge By-Products in Delaware

Inorganic waste constituents[a]	Priority pollutants[b]
Total nitrogen	Volatile compounds
NH$_4$-N	Benzene
NO$_3$-N	Carbon tetrachloride
P	Chloroform
K	Toluene
Ca	Trichloroethylene
Mg	Vinyl chloride
Hg	
Na	Acid compounds
Cu	Pentachlorophenol
Ni	Phenol
Zn	
Pb	Base/neutral compounds
Cd	Hexachlorobenzene
Cr	Phenanthrene
CN	Pyrene
pH	
	Pesticides and PCBs
	Aldrin
	Chlordane
	2,4-D
	Dieldrin
	Heptachlor
	Toxaphene
	Polychlorinated biphenyls (PCB)

Source: Delaware Department of Natural Resources and Environmental Control (DNREC), 1988.
[a] Total analysis required.
[b] Representative examples of each class given; in 1988 there
were 126 priority pollutants identified by the U.S. Environmental Protection Agency.

the actual rate of waste applied either in that year or in total. Typical land-limiting constituents are nutrients, trace elements, and organic compounds. A simplified example of this approach to determine the LLC is illustrated in Table 9-3.

Phosphorus accumulations to excessive levels are common in soils amended with certain types of organic wastes. Most organic wastes used in agricultural land application programs are applied to meet crop N requirements; and because of the N:P ratio of the wastes, more P, is often added than is removed in the harvested portion of the crop. Environmentally harmful levels of soil P occur most frequently when organic wastes are repeatedly applied to the same land for many years, a common practice in situations such as animal-based agriculture, municipal sludge applications in urban areas, and wastewater irrigation from food processing plants. Soil testing results from these areas usually indicate both large percentages of samples in the excessive range for P and soil P values well in excess of the amount required for acceptable crop yields. For example, the latest soil test summaries from the state of Delaware showed that 76% of soil samples from commercial cropland in poultry-producing areas tested in the high or excessive range for P and that 20% of these samples exceeded 135 mg P/kg, relative to a high soil test value of 35 mg P/kg. The long-term effect of this buildup of soil P to excessive levels is the potential for

Table 9-3   Example of the Approach to Determine the Land-Limiting Constituent (LLC) for Land Application of Sewage Sludge

Parameter	Quantity generated (kg/yr)	Site assimilative capacity (kg/ha/yr)	Land area requirement (ha)
Total nitrogen	3000	400	8 (LLC)
Phosphorus	2100	400	5
Cadmium	1.5	0.5	3
Copper	10	14	1
Nickel	25	14	2
Lead	45	56	1
Zinc	160	28	5

Source: Delaware Department of Natural Resources and Environmental Control, (DNREC), 1988.

Note: Assumptions include (1) Site assimilative capacity for N based on crop uptake, ammonia volatilization, and loss in drainage waters with concentrations less than 10 mg NO_3-N/L. (2) Site assimilative capacity for P recognizes the fact that since sludge is applied to meet crop N requirements, excess P (beyond crop requirements) will be applied. Conservation measures are thus required at the site to minimize P losses in runoff, erosion, and drainage. (3) Site assimilative capacity for metals based on maximum cumulative metal-loading rate at the site, assuming a CEC of 10 meq/100 g and a 20-year "site life." Under current Delaware regulations these values are 10, 280, 280, 1120, and 560 kg/ha for Cd, Cu, Ni, Pb, and Zn. Using Cu as an example: Site assimilative capacity = 280 kg/ha ÷ 20 yr = 14 kg/ha/yr. (4) The LLC is defined as the constituent that requires the most land for safe utilization of the sewage sludge, based on the site assimilative capacity. In this case the LLC will be N which requires 8 ha.

environmental problems, particularly in areas where soil erosion and runoff occur near surface waters such as lakes, ponds, and bays. Enrichment of these surface waters with P (and N) can cause eutrophication to occur, resulting in deterioration of surface water quality; in reduced biological diversity; and in extreme conditions, algal blooms and fish kills. There is no simple solution to the P problem associated with organic waste use. Applying organic wastes to meet the P requirements of crops is normally not a viable alternative because the application rates needed would be too low to use all the manure or sludge generated. Larger scale solutions including composting, pelletizing, and transport of organic wastes to nutrient-deficient areas perhaps represent the long-term answer, but require an extensive waste processing and transportation infrastructure that does not exist in many areas.

Trace elements (As, Cd, Cr, Cu, Hg, Ni, Pb, Se, and Zn) are commonly found in many organic wastes, particularly those from industrial or municipal sources. The presence of these elements can dictate both annual and long-term application rates of organic wastes. The most common approach where trace elements are of concern is to establish an annual and total metal loading rate for a site. Although waste applications to meet crop N requirements rarely exceed annual metal loading rates, the cumulative loading rate ultimately eliminates a certain percentage of arable land from further use in organic wasteland application programs. This can be an important issue in urban areas where the availability of land suitable for organic waste application is limited. Animal wastes have not — as of yet — been subjected to the same limitations as municipal and industrial wastes, despite the presence in some manures of trace elements such as As, Cu, and Zn at similar concentrations as those found in sewage sludges. Studies of As, Cu, and Zn in poultry manures have reported concentration ranges of 10–30, 300–1,000, and 200–600 mg/kg for these three elements, respectively, relative to reported median values of 10, 800, and 1,700 mg/kg in municipal sewage sludges from the northeastern U.S.

Organic contaminants can also limit the use of organic wastes as soil amendments. Testing for *"priority pollutants"* (such as pesticides, polychlorinated biphenyls, dioxin, and other industrial contaminants), a standard practice for most municipal sludges and industrial wastes, is now being extended to other organic wastes such as yard waste composts and in certain situations to animal manures (e.g., for pesticides used for insect control in animal production facilities). While clearly necessary, the cost of this testing can be significant. Further, the ability of newer analytical techniques to detect increasingly lower levels of organic compounds (e.g., parts per billion vs parts per million) may exacerbate public concerns about the safety of organic waste use in land application programs.

In the fall of 1992 the U.S. Environmental Protection Agency (EPA) released the "National Sewage Sludge Rule," developed under of the national Clean Water Act for the United States. This rule was developed based on 15 years of comprehensive review of research and management programs using sewage sludge for farms, gardens, forests, and dedicated sites (e.g., landfills). The rule describes general and specific management practices for sewage sludge use in land application and incineration programs. For land application, the general approach is to apply the sewage sludge at an *"agronomic rate"* consistent with crop nutrient requirements. To ensure that this rate does not apply excessive quantities of trace elements or other pollutants, monitoring of sludge composition is required and total quantities of each element that can be applied to a site were established (Table 9-4). For instance, if the rate of sewage sludge needed to meet the N requirement of a corn crop resulted in an annual application of more than 1.9 kg/ha/year of Cd, the sludge application rate would have to be reduced and commercial N fertilizers used to supplement the N provided by the sludge. Cumulative loading rates define the total length of time a site can receive sewage sludge. If a municipality produces a sludge and desires to apply it at rates appropriate for grain crops, and these rates provide 1.0 kg/ha/year of Cd, then the site can be used for 39 years (Table 9-4). The rule also establishes upper limits for the concentration of a trace element to ensure that an excessively contaminated sludge is not applied to soils (e.g., 85 mg/kg for Cd). The rule also states that management practices to control runoff (e.g., buffer zones) be required and that landscapers or homeowners be provided with detailed instructions on the proper means to use sludge-derived products (e.g., sludge composts) for horticultural purposes. Careful recordkeeping, monitoring of sludges for pollutant composition, and practices to ensure that threatened or endangered species are protected are also required under this rule.

The implementation of this rule by the U.S. Environmental Protection Agency is a good example of how our understanding of biogeochemical cycles can be used to develop management programs that identify and prioritize the risk of pollution.

Table 9-4 Maximum Concentrations, Cumulative and Annual
Loading Rate Limits for Trace Elements Present in Sewage Sludge

Trace element	Maximum concentration (mg/kg)	Cumulative load (kg/ha)	Annual maximum load (kg/ha/yr)
Arsenic (As)	75	41	2.0
Cadmium (Cd)	85	39	1.9
Chromium (Cr)	3000	3000	150
Copper (Cu)	4300	1500	75
Lead (Pb)	840	300	15
Mercury (Hg)	57	17	0.85
Molybdenum (Mo)	75	18	0.90
Nickel (Ni)	420	420	21
Selenium (Se)	100	100	5.0
Zinc (Zn)	7500	2800	140

Note: U.S. Environmental Protection Agency, 1992.

Beginning in 1984, the U.S. Environmental Protection Agency reviewed data from studies involving from 200 to 400 pollutants that had been found in sludges. A national research team recommended, based on scientific research on the transformations of these pollutants in the environment (i.e., their biogeochemical cycles), that 50 pollutants be reviewed more intensively. Further evaluation of research and other technical information resulted in the final establishment by the U.S. Environmental Protection Agency in 1992 of national limits for 10 pollutants for land application programs (Table 9-4).

9.2.3 Nutrient Transformations in Wetlands

Wetlands, as defined by the U.S. Fish and Wildlife Service in 1979, are ". . . lands transitional between terrestrial and aquatic ecosystems where the water table is usually at or near the surface or the land is covered by shallow water. . . ." In general, to be classified as a wetland, an area had to meet one or more of the following criteria: (1) predominantly support, at least periodically, hydrophytic (water-loving) vegetation; (2) have, as its substrate, a predominantly undrained *hydric* soil; and (3) have, as its substrate, nonsoil that is saturated or covered with shallow water at some time during the growing season of each year. The U.S. Soil Conservation Service defines hydric soils as ". . . a soil that in its undrained condition is saturated, flooded, or ponded long enough during the growing season to develop anaerobic conditions that favor the growth and regeneration of hydrophytic vegetation. . . ." The criteria for classification of wetlands is under intense scrutiny at the present time because of their value in an undisturbed state and because of their potential value for other land uses once drained (e.g., agriculture, development, mining of peat). There are many types of wetlands, including tidal salt and freshwater marshes, inland freshwater marshes and swamps, peatlands, bogs, prairie "potholes," and riparian (adjacent to rivers) wetlands. Each is a unique ecosystem, but all share a number of vital functions that make wetland preservation a critical environmental issue. Among their more important functions, wetlands reduce erosion, provide control of floodwaters and storm water runoff, maintain water quality by trapping sediments and pollutants, provide wildlife habitats and maintain biodiversity, produce food and timber, and act as an

Table 9-5   Oxidation-Reduction Reactions of Primary Importance in Wetland Soils

Element	Elements or compounds involved in redox reaction		Redox potential for reaction[a] (mv)
	Oxidized species	Reduced species	
Oxygen	Oxygen	H_2O	700–400
	$[0.5O_2 + 2e^- + 2H^+ \rightleftharpoons H_2O]$		
Nitrogen	Nitrate (NO_3^-)	NH_4^+, N_2O, N_2	220
	$[NO_3^- + 2e^- + 2H^+ \rightleftharpoons NO_2^- + H_2O]$		
Manganese	Mn^{4+} (Manganic: MnO_2)	Mn^{2+} (Manganous: MnS)	200
	$[MnO_2 + 2e^- + 4H^+ \rightleftharpoons Mn^{2+} + 2H_2O]$		
Iron	Fe^{3+} (Ferric: $Fe(OH)_3$)	Fe^{2+} (Ferrous: FeS, $Fe(OH)_2$)	120
	$[FeOOH + e^- + 3H^+ \rightleftharpoons Fe^{2+} + 2H_2O]$		
Sulfur	SO_4^{2-} (Sulfate)	S^{2-} (Sulfide: H_2S, FeS)	–75 to –150
	$[SO_4^{2-} + 8H^+ + 7e^- \rightleftharpoons 0.5\ S_2^{2-} + 4H_2O]$		
Carbon	CO_2 (Carbon dioxide)	CH_4 (Methane)	–250 to –350
	$[CO_2 + 8e^- + 8H^+ \rightleftharpoons CH_4 + 2H_2O]$		

[a]   Redox potentials are approximate values and will vary with soil pH and temperature.

aesthetic buffer between urban and industrial areas. Humans alter wetlands by drainage activities, installation of navigational canals, dredging and filling, mining, and point or nonpoint source pollution. It is estimated that between the presettlement era and the 1970s, from 30 to 50% of the wetlands in the U.S. were destroyed.

Biogeochemical cycles in wetlands are dominated by the hydrology of the ecosystem. Under saturated conditions oxygen becomes depleted or diffuses at such a slow rate that anaerobic transformations become the dominant processes in many biogeochemical cycles. From an environmental viewpoint, we seek to use wetlands to reduce pollution of the adjacent aquatic systems. Today many cities, industries, and agricultural enterprises are beginning to investigate the use of constructed wetlands in upland areas to serve as natural wastewater treatment systems. Hence, it is essential to understand the effects of anaerobiosis on chemical and biological reactions that involve pollutants. Because many wetlands do not remain in an anaerobic state year-round, the effects of alternating wet-dry cycles on these reactions can be important as well. Similarly, most wetlands are not anaerobic throughout the soil profile; a shallow aerobic layer often exists in the upper few millimeters.

The factor controlling most of the important reactions in wetlands soils is the *redox potential* of the soil, a measure of the its oxidation-reduction status. In aerobic soils the decomposition of organic matter (oxidation) produces electrons that are then accepted by oxygen, forming water (reduction). When oxygen is absent or its rate of diffusion through the soil is very slow, as in wetland soils or lake sediments, other substances accept these electrons, resulting in the formation of end products other than water (Table 9-5). This was illustrated earlier with nitrate (NO_3-N) (see Chapter 4), where the process of denitrification in wetlands converted a potential aquatic pollutant (NO_3-N) to gaseous N oxides (see Figure 4-7). The sequence of reduction in wetland soils is well understood; hence a knowledge of the system redox potential can be used to predict the dominant electron acceptors present, as shown in Figure 9-4, which illustrates the transformations that occur when a soil is saturated with water. Available O_2 is depleted within one day; NO_3-N becomes the next substrate for electrons produced by anaerobic decomposition of organic matter, followed by Mn oxides and then Fe oxides. As the redox potential (Eh) declines, easily reducible solid forms of Mn disappear and exchangeable Mn^{2+} accumulates, followed by Fe^{2+}

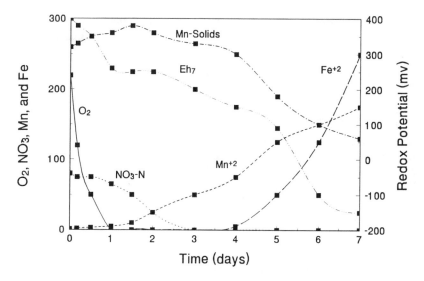

Figure 9-4 Schematic representation of the transformations that can occur when a soil is saturated with water and anoxic conditions develop. (Based on data from Turner and Patrick, 1968.)

as Fe oxides are reduced. Only when all sources of NO_3-N, Mn^{4+}, and Fe^{3+} have been depleted, will the reduction of SO_4^{2-} to sulfide (S^{2-}) occur; this is followed by the anaerobic degradation of organic carbon that results in the production of methane (CH_4), often referred to as "swamp or marsh" gas. Should a soil such as this dry out and aerobic conditions be re-established, the reactions will often reverse; NO_3-N will begin to accumulate as organic matter is mineralized and the soluble Mn and Fe will begin to form insoluble precipitates such as MnO_2 and $Fe(OH)_3$. For an element such as P, where sorption by Fe oxides is an important mechanism of retention in soils, alternating wet and dry cycles can affect both plant uptake and potential for delivery of P to nearby aquatic systems that are sensitive to eutrophication. Research has shown that the development of anaerobic conditions in the lake sediments can reduce Fe oxides and Fe phosphates, resulting in a release of soluble P from these solid phases and increasing the likelihood of eutrophication. Attempts to construct wetlands or riparian zones near agricultural fields to enhance denitrification should, therefore, consider the possibility that this could increase the release of soluble P into streams and rivers bordering these fields.

Other transformations that occur in wetlands, such as the fate of organic compounds (e.g., pesticides, hydrocarbons in storm water runoff) or trace elements, have received less study; hence they are not as well understood as those involving plant nutrients. This will be of particular importance should the use of constructed wetlands as biological filtration zones for urban and industrial wastes increase.

9.2.4. Bioremediation of Soil Contaminated with Organic Chemicals

Bioremediation was defined by the U.S. Environmental Protection Agency in 1991 as "... a process that uses microorganisms to transform harmful substances into nontoxic compounds... one of the most promising technologies for treating chemical

Table 9-6   Classes of Organic Compounds That May Be Suitable for Bioremediation

		Preferred biodegradation process	
Class	Example	Aerobic	Anaerobic
Monochlorinated aromatic compounds	Chlorobenzene	•	
"BTX" (benzene, toluene and xylene)		•	•
Nonhalogenated phenolics and cresols	2-Methyl phenol	•	•
Polynuclear aromatic hydrocarbons	Creosote	•	
Alkanes and alkenes	Fuel oil	•	
PCBs (polychlorinated biphenyls)	Trichlorobiphenyl	•	•
Chlorophenols	Pentachlorophenol	•	•
Nitrogen heterocyclics	Pyridine	•	•
Chlorinated solvents			
Alkanes	Chloroform	•	•
Alkenes	Trichloroethylene	•	•

Source: U.S. Environmental Protection Agency, 1991.

spills and hazardous waste problems." Bioremediation has been successfully used for over 20 years to clean up sites contaminated by spills of oils and other hydrocarbons, and has the potential to degrade many other classes of chemicals (Table 9-6). It is an emerging technology, one that seeks to manipulate the biogeochemical cycling of organic chemicals in a manner that will result in their degradation into harmless by-products such as CO_2 and water. The basic components of a bioremediation process include naturally occurring or genetically engineered microorganisms, a potentially biodegradable pollutant, and a *"bioreactor"* (the location where the bioremediation process takes place). Bioremediation can take place in-situ, in which case the soil volume contaminated by the pollutant is the bioreactor. In-situ bioremediation has been accomplished successfully with both surface soils and with contaminated subsoils and groundwaters, such as those near a leaking underground gasoline storage tank (Figure 9-5). One major advantage of an in-situ approach is that it eliminates the need for excavation and transportation of contaminated soil to another location. Alternatively, the contaminated soil can be removed and taken to another location for aboveground bioremediation by composting, slurrying, or solid-phase treatment in windrows; or even by *"land farming"* (Figure 9-6). In land farming the contaminated soil is spread thinly over cropland; remediation occurs both by biodeg-radation and dilution of the pollutant to extremely low concentrations by mixing with the uncontaminated soil at the site. Land farming has been shown to be an effective technique to remediate soils contaminated by spills of agrichemicals because these materials are already approved for land application and their environmental fates are well-known. In this situation, land farming systems simply represent a means to dilute the contaminant (e.g., a pesticide) to a concentration that allows for effective and timely degradation by soil microorganisms. Regardless of the location, the basic approach involves stimulating the activity of naturally occurring organisms in the soil by aeration, addition of nutrients, and optimization of temperature and moisture. The microorganisms use the carbon in organic pollutants as an energy source and biologically degrade it into less hazardous compounds.

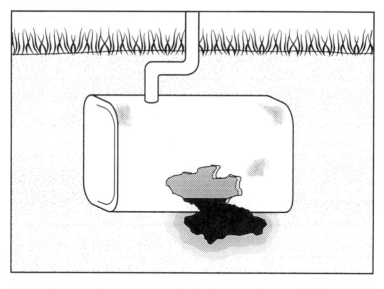

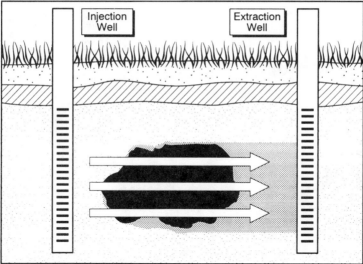

Figure 9-5   Illustration of the process involved in in situ bioremediation of a leaking underground storage tank. (Adapted from U.S. Environmental Protection Agency, 1991.)

As with all efforts to control biogeochemical cycles, bioremediation is not always
a success. Failures can occur when the concentration of the pollutant is high enough
to be toxic to soil microorganisms, or if the degradation process is slower than
another transformation in the biogeochemical cycle that may have an environmental
impact (e.g., leaching). The diversity in properties of potential organic pollutants
means that we may not always understand the optimum environmental conditions to
promote biodegradation, or the toxicity of by-products that result from the process.
Bioremediation can be expensive and require intensive monitoring of soils, leachate,
and air above the site to ensure that the process is proceeding as designed and that

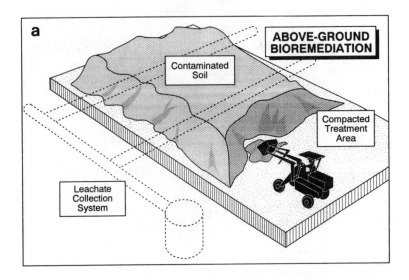

Figure 9-6   Illustration of the processes involved in aboveground bioremediation. (a) Surface bioremediation of contaminated soil. (Adapted from U.S. Environmental Protection Agency, 1991.) (b) "Land farming" of soil contaminated with a pesticide from a spill at an agrichemical facility (Adapted from Felsot, 1991.)

the pollutant or harmful by-products are not escaping by another biogeochemical pathway such as leaching or volatilization.

   Successful bioremediation requires careful planning. If done properly, bioremediation can be a less expensive and an ecologically appealing process. The basic steps in a bioremediation plan include site assessment, preliminary studies on site *"treatability,"* comparison with other alternatives, engineering of the plan, and monitoring of the bioremediation process itself. Site assessment requires an in-depth investigation of the nature of pollutants present, the extent of the pollution, and the current environmental conditions that may limit a bioremediation approach. Preliminary studies can answer important questions on the susceptibility of the pollutants to biodegradation; the presence or absence of other compounds in the soil that may limit microbial activity; and the costs required to provide optimum environmental conditions by aerating, liming, fertilizing, and adjusting soil moisture content. These studies can also determine whether the site should be *"inoculated"* with other microorganisms that are capable of degrading the compound, but are absent from the site or present at very low levels. Finally, the costs and time required for bioremediation must be compared with other alternatives, such as incineration, to determine the most cost-effective process for site cleanup. If bioremediation is the preferred alternative, then an engineering plan is developed, based on preliminary studies, implemented, and carefully monitored by thorough sampling of soil, water, and air at the site.

REFERENCES

Anonymous, Profitable and Sensible Use of Poultry Manure, Special Circular No. 274, Pennsylvania State University, University Park, PA, 1983.

Council on Environmental Quality, *Environmental Trends,* U.S. Government Printing Office, Washington, D.C., 1989.

Delaware Department of Natural Resources and Environmental Control (DNREC), *Guidance and Regulations Governing the Land Treatment of Wastes,* Division of Water Resources, DNREC, Dover, DE, 1988.

Felsot, A., Landfarming can clean up contaminated soil, *Solutions,* February, 16, 1991.

Turner, F. T. and Patrick, W. H., Jr., Chemical changes in waterlogged soils as a result of oxygen depletion, *Trans. IX Int. Congr. Soil Sci.,*4, 53, 1968.

U.S. Department of Agriculture Soil Conservation Service, *Agricultural Waste Management Field Handbook,* No. 651 of the National Engineering Handbook Series, Consolidated Forms and Publications Distribution Center, U.S. Department of Agriculture, Landover, MD, 1992.

U.S. Environmental Protection Agency, *Understanding Bioremediation: A Guidebook for Citizens,* EPA/540/2-91/002, Office of Research and Development, U.S. Environmental Protection Agency, Washington, D.C., 1991.

U.S. Environmental Protection Agency, National Sewage Sludge Rule, 1992.

SUPPLEMENTARY READING

Mitsch, W. J. and Gosselink, J. G., *Wetlands,* Van Nostrand Reinhold, New York, 1986.

Overcash, M. R., Humenik, F. J., and Miner, J. R., Introduction to livestock waste management, in *CRC Livestock Waste Management,* Vol. I, Overcash et al., Eds., CRC Press, Boca Raton, FL., 1983.

10 THE ATMOSPHERE: GLOBAL CLIMATE CHANGE AND ACID PRECIPITATION

A description of the atmosphere was provided in Chapter 2. This chapter discusses two important phenomena related to anthropogenic changes in the atmosphere: global climate change and acid precipitation. Each has numerous effects and inter-relationships with soils.

10.1 GLOBAL CLIMATE CHANGE

Global warming refers to the possibility that the global average air temperature may be increasing because the composition of the concentrations of various gases in the atmosphere are changing. These gases create the greenhouse effect, which allows the atmosphere to trap radiant energy that would otherwise radiate freely away from the earth. The greenhouse effect allows life as we know it to exist. Without any greenhouse effect the surface of the earth would be 33°C colder than it is now. Unfortunately, climatic changes are likely to occur as the mean global temperature increases. These changes are as yet relatively unpredictable beyond the knowledge that an increased global temperature will influence the driving forces for our weather. The changes in the climate may cause increased variability in temperatures and precipitation. Extended droughts and hot spells may become more common. Both the total annual precipitation and the distribution of precipitation during the year would be subject to change, which could dramatically affect where and how food is produced.

A simple example of the greenhouse effect can be found in an automobile parked outside on a sunny day with the windows closed. Solar radiation passes through the glass and warms the interior surfaces of the automobile. These surfaces then radiate that heat into the air inside the automobile, but the glass prevents the heated air and emitted radiation from escaping, and the temperature inside the automobile rises. The same phenomenon occurs in the earth's atmosphere. Incoming solar radiation warms the surface of the earth and the heat is radiated back into the atmosphere. The greenhouse gases act as the glass in our automobile example and prevent some of the heat from escaping the atmosphere. The higher the concentration of the greenhouse gases, the more heat is trapped. Incoming solar radiation has sufficient energy to pass through the atmosphere. The reemitted infrared radiation — measured as heat — is

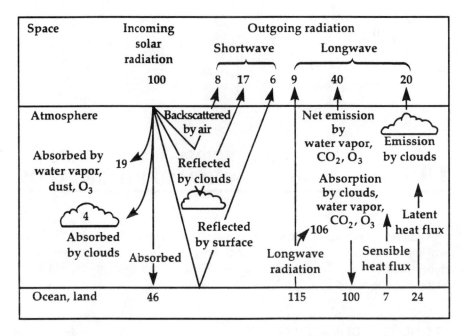

Figure 10-1 The global energy balance in units of percent of incoming solar radiation. (From MacCracken and Luther, 1985.)

absorbed by greenhouse gases, and the energy becomes trapped in the atmosphere. The gases that absorb the infrared radiation are called *radiatively active.*

The greenhouse effect can be shown from the global energy balance depicted in Figure 10-1. From the incoming solar radiation 100 units enter the atmosphere and 31 units are reflected to space. Of the remaining units, 46 are absorbed by the surface and are partly used to directly heat the atmosphere (sensible heat flux) and to evaporate water (latent heat flux). The fact that 100 units of infrared radiation are emitted downward from the atmosphere to the surface from absorption by clouds, water vapor, carbon dioxide (CO_2), and ozone (O_3), in addition to the 46 units absorbed by the surface directly from incoming solar radiation, allows the surface to warm more than it would by incoming solar radiation alone. The anthropogenic greenhouse effect causes an increase in the flux of infrared radiation downward to the surface.

That the mean global temperature is, in fact, increasing has yet to be conclusively shown. The actual determination of the average global temperature is controversial in itself. The mean global temperature is the mean of the mean annual temperature for a number of locations spread across the earth. The mean annual temperature for a given location is the mean of the high and low temperatures for each day in the year. Experts disagree on how a representative measurement can be obtained without being biased by heat radiated from metropolitan areas or by the lack of observations over the oceans, which cover the majority of the earth's surface. Despite the disagreements, it is generally believed that the average global temperature has increased 0.4–1.0°C over the last 100 years. The trend of increasing temperature is shown in Figure

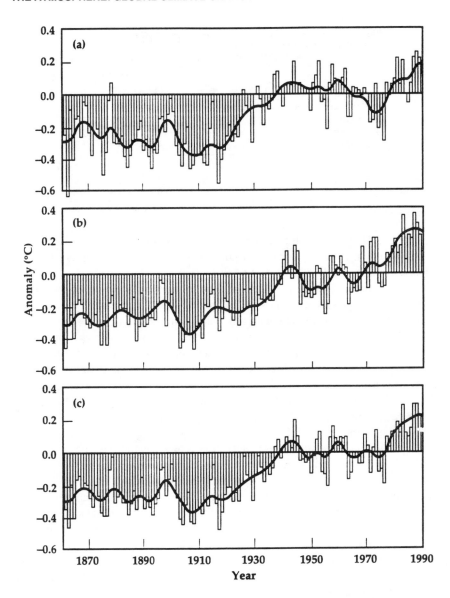

Figure 10-2 Combined land-air and sea surface temperatures for the period 1861–1989 for the northern hemisphere (a), southern hemisphere (b), and the global mean (c). The units of the y-axis are the departure from the average for the period 1951–1980. (From Wuebbles and Edwards 1991.)

10-2. It is known that the six warmest years on record occurred in the 1980s and 1990. Still the observations on temperature are within the natural variability of the climate, and there is further disagreement on whether the trends of the past two decades will persist. By the guidelines of the scientific method, discussed in Chapter 1, one could not conclude that global warming is real yet. This fact has been used by some to justify a lack of actions aimed at reducing the loadings of greenhouse gases to the

Table 10-1   Predicted Changes in Annual Temperature and
Precipitation for Three U.S. Regions by Three Global Circulation
Models Assuming a Doubling of Atmospheric CO_2 Concentrations

	Model		
	A	B	C
Potential change in annual temperature (°C)			
Great Lakes	4.5	6.2	3.4
Great Plains	4.5	5.0	3.4
Southeast	4.6	4.6	3.5
Potential change in annual precipitation (cm)			
Great Lakes	4.1	1.6	3.7
Great Plains	−9.1	2.3	1.8
Southeast	9.1	−3.0	2.2

Source: White, 1990.

atmosphere. It is also difficult to appreciate the potential change in our climate due to a change of only a few degrees. To put this into perspective, consider that the climate has warmed only 5°C since the last ice age.

Global circulation models (GCMs) are used to predict the effects of increasing concentrations of greenhouse gases on components of the climate. Typically, the models will compare the climate with a doubled atmospheric CO_2 concentration, a condition that will likely occur before the middle of the next century. Table 10-1 presents predicted changes in annual temperature and precipitation with a doubling of CO_2 in the atmosphere from three GCMs for three regions in the United States. The models are consistent in that all predict at least a 3.4°C increase in mean temperature. The models are not as consistent with predictions for annual precipitation with an 11.4-cm range from the models for the Great Plains region. In light of a 3.4–5.0°C increase in temperature, an 11.4-cm variation in annual precipitation becomes important. The information also illustrates that the effects of any anthropogenic greenhouse effect will vary by region. A great deal of uncertainty exists in the area of climate modeling.

It is known that the concentrations of the five most important anthropogenic greenhouse gases (carbon dioxide, methane, nitrous oxide, ozone, and chlorofluoro-carbons) have increased substantially since preindustrial times. These concentrations are given in Table 10-2 along with some general information about each gas. The potency of each greenhouse gas is expressed as its radiative forcing relative to CO_2. Generally speaking, the radiative forcing value represents the capacity of a single molecule of a greenhouse gas to absorb infrared radiation compared to one molecule of CO_2. Methane (CH_4), nitrous oxide (N_2O), chlorofluorocarbons (CFCs), and O_3 are all more efficient absorbers than CO_2; and this is why they contribute significant proportions to the anthropogenic greenhouse effect at relatively low concentrations.

Carbon dioxide is present in the atmosphere in the highest concentration, by a factor of at least 200, compared to the remaining greenhouse gases; and therefore has the highest relative contribution to the anthropogenic greenhouse effect. It contributes approximately 60% to the greenhouse effect. It is the least potent of the greenhouse gases, however, with its contribution to global warming being primarily due to the large quantities present. The primary sources of CO_2 are the combustion of fossil fuels and deforestation. Methane contributes approximately 15% to the greenhouse effect. It is a product of anaerobic metabolism of C compounds. Such

Table 10-2 Characteristics of the Five Primary Greenhouse Gases

	Carbon dioxide (CO_2)	Methane (CH_4)	Nitrous oxide (N_2O)	Chlorofluorocarbons (CFCs)	Tropospheric ozone (O_3)
Principal natural sources	Release from land and oceans	Wetlands, enteric fermentation (non-domestic animals)	Release from land and oceans	None	Some transport of O_3 from stratosphere, some photochemical production in troposphere
Principal anthropogenic sources	Fossil fuel use, land-use conversion cement manufacturing	Rice culture, enteric fermentation (domestic animals), biomass burning, natural gas losses	Fertilized soils, fossil fuel use, biomass burning	Refrigerants, foams aerosol propellants	Reactions with NO_2, CH_4, and hydrocarbons
Present atmospheric concentration (parts per billion by volume)	353,000	1720	310	CFC-11: 0.28 CFC-12: 0.48 Others[a]: 0.005–0.12	20–100
Preindustrial atmospheric concentration (parts per billion by volume)	280,000	790	280–290	0	10
Atmospheric lifetime	50–200 yr	10 yr	120–150 yr	CFC-11: 65 yr CFC-12: 120 yr Others: 0–400 yr	hr to days
Radiative forcing relative to an equal concentration (volumetric basis) of CO_2	1	58	206	CFC-11: 3970 CFC-12: 5750 Others: 3710–5440	—[b]
Estimated relative contribution to the anthropogenic greenhouse effect	60%	15%	5%	12%	8%

Source: Wuebbles and Edwards, 1991; Earthquest, 1990.

a Includes CFC-113, CFC-115, and HCFC-22.

b Complex dependence as a function of altitude.

conditions are found in flooded soils and in the stomachs of ruminant animals. Nitrous oxide contributes approximately 5% to the greenhouse effect. It is produced as a result of the use of N fertilizers, land use conversion, and combustion of fossil fuels. The CFCs are a group of gases that contribute 12% to the greenhouse effect. They are the most potent of the greenhouse gases with atmospheric concentrations less than 1 ppb, by far the lowest of all of the greenhouse gases, and with the third largest contribution to the effect. Finally, O_3 in the troposphere is believed to be responsible for 8% of the anthropogenic greenhouse effect. Ozone is produced through a number of mechanisms associated with urban environments and is a pollutant in the lower atmosphere. Water vapor is also a very important radiatively active gas, although the concentrations of water vapor in the atmosphere generally do not change as a direct result of human activities. However, a warmer climate will induce greater evaporation worldwide, and the additional water vapor will likely contribute to additional warming.

Soils can be a source or sink for CO_2, CH_4, and N_2O; and these gases will be discussed in this chapter. Soils play no role in the cycling of O_3 or CFCs, and comments on these two gases will be very brief. The CFCs — used primarily as refrigerants and in aerosols — are unique in that they are a class of compounds for which input to the atmosphere can be eliminated, provided less harmful substitutes are developed. This is an important point since CFCs are such potent greenhouse gases and have also been shown to participate in the destruction of O_3 in the upper atmosphere. Since O_3 is a pollutant in the lower atmosphere, efforts are already underway to reduce O_3 inputs to the atmosphere which will have beneficial effects toward reducing global warming.

10.1.1 Carbon Dioxide

A more thorough appreciation of why CO_2 concentrations are increasing in the atmosphere can be obtained by examining the global C cycle (Figure 10-3). For the time period immediately prior to large-scale population of the earth, atmospheric CO_2 concentrations were essentially constant. All biomes across the earth were essentially at the state of climax vegetation, and CO_2 released by the decomposition of dead plant material was balanced with CO_2 uptake by actively growing vegetation. All other C pools were relatively stable. The influence of humans has essentially been to take C stored in recoverable fossil fuels and in forests and place it in the atmosphere, after it has been converted to CO_2, at a faster rate than it can be removed by plant growth or by absorption into the oceans. Atmospheric CO_2 concentrations, as determined from air samples (since about 1950) and from air trapped in ice cores (pre-1950) for the last 250 years, are shown in Figure 10-4.

A summation of the C transfer rates in Figure 10-3 shows that the net transfer of C into the atmosphere is 3–5 Gt/year (Gt = 10^9 metric tons). The 6–8 Gt/year placed into the atmosphere from combustion of fossil fuels and deforestation is slightly offset by a net transfer of C into the oceans at 3 Gt/year. Humans have some control over three processes in the C cycle: deforestation, combustion of fossil fuels, and to a lesser extent the amount of C taken up by growing plants. Carbon used for plant growth is stored both aboveground and in the soil while the plant is alive. Some of the C remains in the soil as organic matter after the plant dies and decomposes.

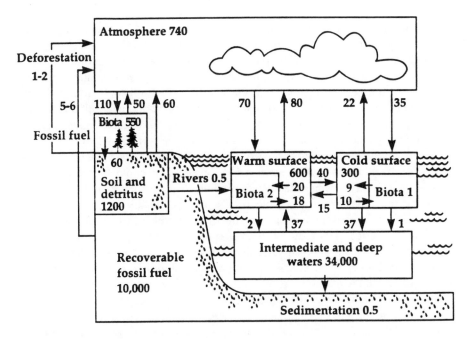

Figure 10-3 The global C cycle (in Gt/year). (From Moore, 1988.)

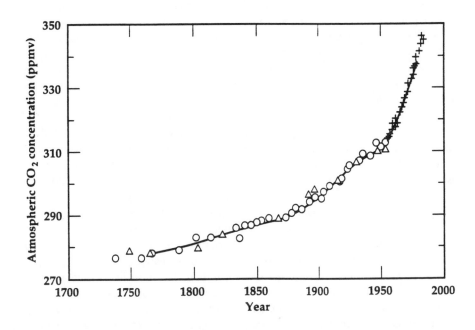

Figure 10-4 Atmospheric CO_2 concentrations from preindustrial times to present as determined from air trapped in ice cores and actual air samples. (From Lashof and Tirpak, 1989.)

The role of soils in the global cycling of C is quite diverse. First, soil and detritus contain 1200 Gt of C, which is one of the larger pools of C in the global C cycle. Plants fix atmospheric CO_2 as they grow. Soil microorganisms decompose dead plants, returning a portion of the C to the atmosphere as CO_2 and retaining a portion of the C as soil organic matter. Clearly, the significance of soils in C storage should not be overlooked. In fact, when soils are taken out of natural vegetation and used for agricultural production, they will typically lose 25–50% of their organic matter, unless properly managed. When soil erosion occurs, there is also a loss of C in the eroded material. Increasing soil organic matter contents has been suggested as a means of offsetting rising atmospheric CO_2 levels. This could be accomplished through animal waste management (returning all animal waste to soils), reducing tillage intensity (conserves soil organic matter), minimizing fallowing, planting more grasses and deep-rooted legumes, and maximizing the return of crop residues to soils. All of these practices help reduce soil erosion in addition to increasing the amount of C returned to the soil. It is estimated that 63 million metric tons of additional C/yr could be stored in cultivated U.S. soils alone, which is approximately 1.5% of the net transfer of C into the atmosphere.

Energy consumption by agriculture is another area of concern with regard to CO_2 contributions to the atmosphere. Food production requires direct energy inputs from various fuels used for machinery and drying of crops, and energy used indirectly from the manufacture of pesticides and fertilizers and the generation of electricity. Estimated C emissions from energy consumption from U.S. agriculture are given in Table 10-3. Fertilizer manufacturing produces the majority of C emissions, representing 30% of the total. The reduction in C emissions from 1974 to 1987 reflects the conversion from gasoline to more efficient diesel fueled equipment. Total C emissions from U.S. agriculture are less than 1.0% of the net transfer of C into the atmosphere. While this may seem insignificant from a global view, the importance becomes evident if one considers that the trend is toward agriculture across the globe to become as energy intensive as U.S. agriculture. Fertilizer use, for example, has increased sharply in Asia and the former U.S.S.R. during the last 25 years. From this perspective, efforts to reduce energy use in food production seem prudent.

Overall, strategies for dealing with increasing CO_2 levels in the atmosphere follow the lines of reducing CO_2 inputs to the atmosphere or increasing the amount of C stored in various C pools. For the former, efforts to increase fuel efficiency, increase use of nonfossil fuel based energy sources (e.g., wind, solar, nuclear), prevention of soil erosion, and slowing deforestation will all directly reduce C emissions. For the latter, revegetation (forests and grasslands), increasing soil C levels, and promotion of beneficial reuse of organic waste materials (manures, sludges, etc.) have the potential to increase the amount of stored C. The reader is reminded of the magnitude of amounts of C that need to be dealt with. Significant changes in the size of any C pool will require some degree of global cooperation. The most likely outcome will be a decrease in the rate of increase of CO_2 levels in the atmosphere. The easiest actions to implement are the "no regret" type of actions. These reflect changes that should be made regardless of the global warming problem. There are a number of reasons, for example, that soil erosion should be prevented or that fuel efficiencies

Table 10-3   Estimated C Emissions from Energy
Consumption by U.S. Agriculture in 1974 and 1987

Energy source	C emissions (millions of tons)	
	1974	1987
Direct energy		
Gasoline	10.3	4.2
Diesel	8.0	8.9
LP gas	2.0	0.9
Natural gas	2.2	0.9
Invested energy		
Electricity	5.5	6.1
Fertilizer	9.5	9.9
Pesticides	1.8	2.4
Total	39.3	33.2

Source: Bird and Strange, 1992.

should be improved. If global warming happens to be less serious than is now believed, the resources spent on those actions would not be wasted.

The difficulties in predicting the effects of global warming on climate have been discussed briefly. Adding to the complexity of the situation are the potential responses of plants themselves to the increased CO_2 concentrations, without regard to the effects of changing climate on plant productivity.

Most plants fall into one of two general categories based on differences in their biochemical pathways for fixing atmospheric CO_2; C_3 plants (e.g., wheat, soybeans, cotton) take CO_2 from the air into mesophyll cells where the Calvin cycle converts it to sucrose; C_4 plants (e.g., corn, sorghum, millet) also take CO_2 from the air into mesophyll cells but the fixed CO_2 (as a C_4 molecule) is transported to the bundle sheath cells, where the Calvin cycle takes place. This "pumping" of CO_2 from the mesophyll cells into the bundle sheath cells maintains a greater CO_2 concentration gradient between the atmosphere and the mesophyll cells, which increases the ability of the C_4 plant to assimilate CO_2. Thus, C_4 plants are more efficient users of CO_2.

The C_3 plants, being less efficient users of CO_2, will actually increase their photosynthetic rate in response to elevated CO_2 concentrations, provided that other growth factors are not limiting. Thus, one could predict increased growth from C_3 plants as atmospheric CO_2 levels increase. The C_4 plants will also respond positively to increased CO_2 concentrations, although not to the extent that C_3 plants would. Both C_3 and C_4 plants tend to use water more efficiently when CO_2 levels are higher. This is because the stomata (openings in leaves that provide the means of gas and water exchange between plants and the atmosphere) do not open as much in the CO_2-enriched environment compared to current conditions, and less water is lost during photosynthesis. This effect is more pronounced with C_4 plants. In the case of C_3 plants, there is also an increase in plant growth without an increase in water use. Both C_3 and C_4 plants can be more efficient in their use of nutrients, especially N, which is a positive response.

Unfortunately, water and nutrients are already the primary limiting factors for plant growth across much of the earth, and this will likely remain the case as the climate changes. Clearly, if large-scale increases in plant productivity were possible due to increasing atmospheric CO_2 levels, then atmospheric CO_2 concentrations

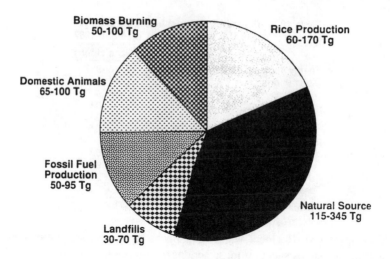

Figure 10-5 Current sources of global CH_4 emissions. (From Burke and Lashof, 1990. With permission.)

would not be as high as they are today. Increased biomass production has, however, likely reduced the magnitude of increase in atmospheric CO_2 concentrations. Water may even become more of a limiting factor in light of increasing temperatures.

10.1.2 Methane

Methane is produced by methanogenic bacteria anytime organic matter decomposes anaerobically. These conditions are typically found in flooded soils; in the digestive systems of ruminant animals; and in situations where organic wastes are stored or handled as liquids, as is often done with manures and sewage sludges. Humans have increased CH_4 concentrations in the atmosphere by increasing the amount of C that is metabolized anaerobically. Similar to CO_2, atmospheric CH_4 concentrations were relatively constant prior to large-scale population of the earth. Wetlands and native ruminant animals were the primary sources of CH_4 at that time. Today, rice culture, domestic ruminant animals, landfills, and our own use of CH_4 as a fuel source provide additional CH_4 to the atmosphere. Figure 10-5 shows a breakdown of current global annual emissions of CH_4 by source. Of all the greenhouse gases, CH_4 emissions are the most closely tied with the production of food on a global basis. It is estimated that 40% of global CH_4 emissions are from rice culture and from the digestive processes and waste handling associated with domesticated livestock alone.

Methane is a more potent greenhouse gas than CO_2. With an atmospheric concentration of approximately 1700 ppb, it contributes 15% to the anthropogenic greenhouse effect. The unique thing about CH_4 is the short atmospheric lifetime of 10 years compared to CO_2, N_2O, or CFCs (Table 10-2). The fate of atmospheric CH_4 is not completely understood. It is known that soil bacteria responsible for nitrification can oxidize CH_4 as well as NH_4^+. Therefore, aerated low N soils can be a CH_4 sink. Most

CH_4 is dissipated via reaction with tropospheric OH, however. The short atmospheric lifetime of CH_4 also suggests that efforts to reduce CH_4 emissions could be effective in slowing the increase in the rate of global warming.

Efforts to directly reduce CH_4 emissions resulting from food production will not be easily implemented because of increasing population and demand for food in areas of the world where rice is a major staple. In addition, large sectors of the world population have low quality diets, and historically diet improvement has come about by increasing the proportion of meat in the diet. This would suggest increasing numbers of domestic animals in some areas of the world. Research may be able to reduce the amount of CH_4 produced per unit of rice produced. This could be accomplished by breeding rice with a higher grain:straw ratio, which results in less straw available for anaerobic decomposition; or by increasing the proportion of upland rice varieties grown. Other water-use strategies may reduce CH_4 emissions from rice production. Research may also reduce CH_4 emissions from livestock, particularly ruminants. The CH_4 emitted from these animals reflects a loss of feed energy and is an inefficiency. Efforts to increase feed efficiency or to increase the rate of weight gain by livestock will decrease the amount of CH_4 emitted.

Direct efforts at reducing CH_4 emissions will need to focus on natural gas distribution systems, fossil fuel production methods, landfills, and other waste handling methods. It has been estimated that some natural gas distribution systems lose as much as 3% of the gas through leakage. Oil exploration and coal mining also release large amounts of CH_4, as do landfills and some organic waste handling systems. In each of these examples there are preventable losses of CH_4, and methods for capturing or retaining CH_4 will benefit the global climate problem.

10.1.3 Nitrous Oxide

Nitrous oxide is produced in soils through both biotic and abiotic means. Various microbial pathways produce N_2O, including nitrification and denitrification. Nitrous oxide is released from all soils, regardless of cultivation or fertilization, although N fertilization will increase N_2O emissions because of the increase in the amount of substrate available for nitrification and denitrification. Emissions of N_2O from soils can be stimulated by land use conversion, particularly the conversion of land from any native vegetation to agricultural production. As discussed previously, soil organic matter levels tend to decrease when soils are first cultivated. Some of the N in the organic matter will be lost as N_2O. Nitrous oxide is also produced when reduced N compounds in fossil fuels or other biomass materials are oxidized during combustion. As a result of increases in the activities described above, the concentration of N_2O in the atmosphere has increased compared to preindustrial times.

Nitrous oxide is a more potent greenhouse gas than CO_2 or CH_4, although it contributes only 5% to the anthropogenic greenhouse effect. Means for reducing CO_2 emissions — such as reducing fossil fuel use and conserving soil organic matter (that have already been described) — will also reduce N_2O emissions since N is an integral component of fossil fuels, biomass, and soil organic C. Direct efforts toward reducing N_2O emissions will have to address fertilizer N use. Fertilizer-derived N_2O emission is a complicated process that can be influenced by factors such as fertilizer type, N rate, application method, tillage, soil moisture, temperature, soil organic C content, and microbial activity. Overall, a reduction in fertilizer-derived N_2O emissions will require increases in N use efficiency, which will occur to some extent with elevated atmospheric CO_2 levels. Methods for increasing N use efficiency were discussed in Chapter 4. Presently, fertilizer-derived N_2O emissions are estimated to account for 2.5% of global N_2O emissions, assuming that approximately 1.0% of fertilizer N is lost as N_2O. Nitrous oxide emissions due to fertilizer use would be expected to increase worldwide since N fertilizer use is increasing, particularly in countries with high population densities and increasing food demands.

10.1.4 Uncertainties and Complexities

Within the general topic of climate change the unknown far outweighs the known. Beginning with the climate itself, it is generally accepted that the climate will change although the nature of the changes are largely unpredictable. One of the reasons for the uncertainty is the large number of feedback mechanisms. That is, a change in one mechanism will induce additional changes in others such that the net effect of the first change becomes difficult to predict. A simple example was given earlier when it was noted that we know the greenhouse gases are radiatively active and warm the atmosphere, but it cannot be said with certainty that continued increases in the concentrations of the gases will cause a continued proportional increase in warming. It is useful to discuss these feedbacks because they influence some of the proposed actions for responding to global climate change. It also assists in appreciating the complexity of the global warming problem. Bear in mind that feedbacks can be both positive and negative. Several examples will be offered to illustrate this complexity.

Undoubtedly water vapor will play a role in potential climate changes. As the climate warms there will be more evaporation because more sensible heat is available and warmer air can hold more water than cooler air. The higher concentration of water vapor in the atmosphere may produce more cloud cover. If that cloud cover occurs frequently as a uniform layer (stratus type) at low altitudes over large areas, the clouds may have a net cooling effect. If that cloud cover occurs frequently as a broken layer (cumulus type), the additional water vapor will likely absorb heat and contribute to warming. The additional water vapor will also increase the hydroxyl concentration in the atmosphere. The hydroxyls can react with both CH_4 and O_3, reducing their concentrations and acting as a positive feedback for atmospheric warming.

The increasing temperature will have a multitude of effects. Warmer soils will induce oxidation of C, and the total amount of C stored in soils may decrease and further increase atmospheric CO_2 concentrations. Warmer air temperatures may also allow the northward expansion of forested regions. Provided that the aforestation is not accompanied by an equivalent amount of deforestation elsewhere, this could have a positive impact on CO_2 concentrations in the atmosphere. Warmer oceans may support higher phytoplankton populations, which absorb CO_2, but warmer water holds less CO_2 itself. The warming of the oceans lessens the increase in temperature of the atmosphere.

One aspect of global climate change that is often forgotten is the potential adaptability of the population to the changing climate. The changes are sometimes considered in a catastrophic context; productive land suddenly transformed to a desertlike state. Climate change will occur more gradually, and the population and food production practices will have to adapt. The growing of grain crops may need to take place at higher latitudes in the northern hemisphere. Canada and northern areas of the former U.S.S.R. and China could become more important grain producing areas than they are today. The shift in production would occur gradually over a number of years or even decades. In addition, crop breeding and biotechnology may produce corps that are more heat and drought resistant. These points are not intended to downplay the significant economic and social issues associated with a shifting agriculture, but instead to illustrate that adaptations are possible.

As the net effects of the increase in concentrations of the greenhouse gases in the atmosphere are impossible to predict at this time, so are the net effects of any actions taken to reduce emissions of any of the gases into the atmosphere. Most experts would agree that significant climatic changes have already been induced, although we do not know what they are yet. This was illustrated in Table 10-1. It is fairly certain that the climate has or will warm as a whole and that the resulting net effects will not likely be favorable. Therefore, any realistic efforts made to reduce emissions of greenhouse gases would seem worthwhile regardless of the uncertainties.

10.2 ACIDIC DEPOSITION

One of the most important, highly publicized, and controversial aspects of atmospheric pollution has been the issue of *acid rain*. To many, the term evokes frightening images of industrial pollution literally raining down on the earth and causing serious and perhaps irreversible environmental damage. Technically, the term *acidic deposition* — which includes not only rainfall, but also acidic fogs, mists, snowmelt, gases, and dry particulate matter — is a more precise description of the problem. The primary origin of acidic deposition is the emission of sulfur (SO_2) and nitrogen oxides (NO_x) when fossil fuels are burned for energy production. Typical sources of these airborne pollutants include coal and oil-burning electric power plants, automobiles and other vehicles, and large industrial operations (e.g., smelters) (Figure 10-6). Once these gases enter the earth's atmosphere they react very rapidly with moisture in the air to form sulfuric (H_2SO_4) and nitric (HNO_3) acids. The pH of natural rainfall in equilibrium with atmospheric CO_2 is about 5.6; however, as shown in Figure 10-7, for the U.S., the pH of rainfall is less than 4.5 in many industrialized areas of the world. When these acids are then deposited in lakes, streams, forests, and other natural ecosystems, either by wet or dry deposition, a number of adverse environmental effects are believed to occur. Among the most serious are direct damages to vegetation, particularly forests, and changes in soil and surface water chemistry that can adversely affect plants and animals.

Clearly, emissions of sulfur dioxide (SO_2) and NO_x have increased in the 20th century (Table 10-4); however, as mentioned earlier with regard to other atmospheric pollutants, there is a fair degree of scientific uncertainty as to the actual means by

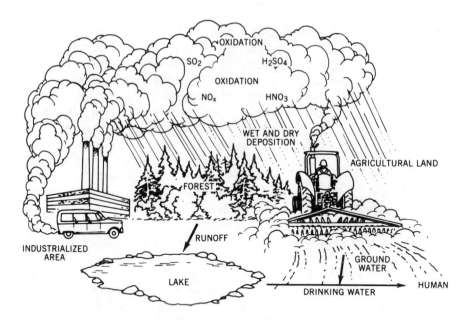

Figure 10-6 Generalized process of production and distribution of acidic deposition. (Adapted from U.S. Environmental Protection Agency, 1980.)

which acidic deposition affects our environment. For example, to quote a 1986 report of the National Research Council, "…data uncertainties make it difficult to reach unambiguous conclusions about trends in alkalinity and acidity in Adirondack lakes during the past 50 years. The weight of chemical and biological evidence indicates that atmospheric deposition of sulfate has caused some Adirondack lakes to decrease in alkalinity"; to quote from the same report, with reference to eastern U.S. forests, "…current data do not permit adequate evaluation of the roles of acidic deposition and other factors in this phenomenon." These quotes reflect the relatively common reluctance of many scientists to overgeneralize cause and effect relationships in an extremely complex environmental problem. A major cause of this uncertainty is the fact that other changes in atmospheric chemistry and global climate change have paralleled the increase in acidic deposition.

Most industrialized countries, however, have concluded that the issue is serious enough to proceed without final resolution of all scientific issues, a process that could take decades. For example, by 1988, 21 European nations had signed a "Long-Range Transboundary Air Pollution" agreement that had target reductions in SO_2 emissions of 30% by 1993 from the 1980 baseline. Progress is sure to be slow because of the lack of alternative energy sources to those that produce SO_2 and NO_x and the billions of dollars needed to further reduce emissions of these gases from current sources. Unfortunately, as with many other environmental issues, solutions to the acidic deposition problem are complex, expensive, and long term in nature.

In this section we will address the sources and geographic distribution of acidic deposition, describe the most common types of problems it can produce, and give an overview of some remediation efforts currently used to reverse the effects of acidification.

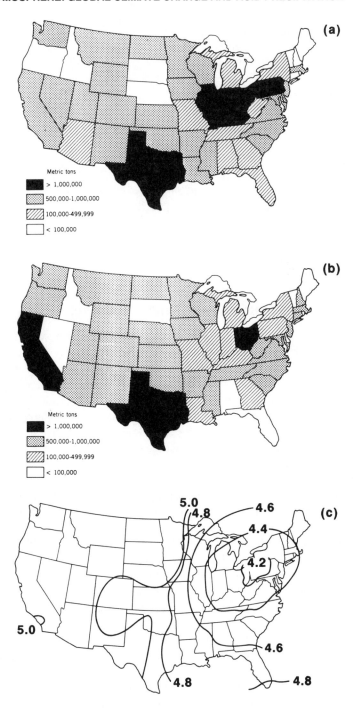

Figure 10-7 Geographic distribution of (a) sulfur dioxide emissions, (b) nitrogen oxide emissions, and (c) acidity of precipitation in the U.S. (Adapted from Council on Environmental Quality, 1989; U. S. Environmental Protection Agency, 1988; and World Resources Institute, 1988.)

Table 10-4   Historical Trends for the Emissions of Sulfur and Nitrogen Oxides

Region	1900	1925	1950	1970	1985
	Sulfur dioxide emissions (million Mg/yr)				
United States					
Contiguous[a]	3.2–5.5	8.2–13.2	8.0–12.7	11.3–18.2	8.9–13.2
Northeast	1.1	3.5	2.0	2.6	1.6
Southeast	0.6	1.4	1.6	3.2	2.8
Industrial midwest	1.8	3.9	3.5	4.8	3.6
Upper midwest	0.4	0.9	0.7	0.9	0.5
Southeastern Canada	0.2	0.9	1.8	3.2	1.7
United Kingdom	1.4	—	2.3	3.1	1.8
	Nitrogen oxide emissions (million Mg/yr)				
United States					
Northeast	0.1	0.9	1.9	3.9	3.5
Southeast	0.1	0.4	1.6	5.6	6.4
United Kingdom	—	—	0.2	0.7	0.7

Source: Charles, 1991 and Mason, 1992.
[a] Values are low and high estimates for each year.

10.2.1 Sources and Distribution of Acidic Deposition

The nature of acidic deposition, that is the relative percentages of H_2SO_4 and HNO_3 present, is controlled largely by the geographic distribution of the sources of SO_2 and NO_x. In the U.S., H_2SO_4 is the main source of acidity in precipitation in the midwest and northeast because coal-burning electric utilities in these states and Canada emit large quantities of SO_2. In the western U.S., utilities and industry burn coal with a lower S content; and HNO_3 is of more concern, particularly in densely populated areas such as California where cars and other vehicles that burn gasoline are major sources of NO_x. The geographic nature of SO_2 and NO_x emissions and the acidity of precipitation in the U.S. are clearly shown in Figure 10-7.

The issue of acidic deposition crosses state and national boundaries because, although acidic compounds can be deposited short distances from the source of generation, they may also be transported hundreds of miles before being returned to the earth in rainfall or other forms. It has been estimated, for example, that 30–40% of the sulfur deposited in the northeastern U.S. originated in the industrial midwest. On an international scale, the U.S. and Canada have debated the sources of acidic deposition and the economics of reversing acidification in eastern North America for years and have yet to develop a comprehensive strategy to address the problem. Similarly, in Europe, the small size of many countries means that emissions in one industrialized area can readily affect forests and lakes in another nation. One recent study estimated that 17% of the acidic deposition falling on Norway originated in Britain and that 20% in Sweden came from eastern Europe. Recent political changes and ongoing economic instability in eastern European countries that burn high sulfur coal for energy generation may mean even more delays in implementing international agreements to reduce emissions of acidic compounds.

10.2.2 Environmental Effects of Acidic Deposition

Acidic deposition can have many environmental impacts. Prior to discussing the mechanisms involved and some approaches to reverse the effects of acidification, it is important to first briefly review the nature of acidity in the soil and aquatic environments. In simplest terms, acids are substances that tend to donate protons (hydrogen ions, H^+) to another substance in a chemical reaction. Acids are often classified as strong or weak, with strong acids tending to completely dissociate (lose H^+) in water and weak acids undergoing only partial dissociation. In addition to H_2SO_4 and HNO_3, other common strong acids are hydrochloric (HCl) and phosphoric (H_3PO_4); weak acids include carbonic (H_2CO_3), acetic (CH_3COOH), and boric (H_3BO_3) (note that only H_2SO_4 and HNO_3 are important components of acidic deposition). The term pH is used to indicate the relative acidity of a solution (or soil) and is defined as the negative logarithm of the activity of the H^+ in solution [pH = $-\log$ (H^+)]; hence the greater the H^+ concentration in a solution, the lower the pH. Pure water has a pH of 7.0, natural rainfall about 5.6, and severely acidic deposition <4.0. The pH of most soils ranges from 3.0 to 8.0; for most crop production systems, recommended pH values range from 6.0 to 7.0. When acids are added to soils or waters, the decrease in pH that occurs depends greatly on how *buffered* the system is. Buffering refers to the ability of a system to maintain its present pH by neutralizing added acidity. Clays, organic matter, oxides of Al and Fe, and Ca and Mg carbonates (limestones) are the components responsible for pH buffering in most soils. Acidic deposition, therefore, will have a greater impact on sandy, low organic matter soils than those higher in clay, organic matter, and carbonates. In fresh waters, the primary buffering mechanism is the reaction of dissolved bicarbonate ions with H^+, i.e.:

$$H^+ + HCO_3^- \rightleftharpoons H_2O + CO_2 \qquad (10\text{-}1)$$

In many natural systems, the damage from acidification is often not directly due to the presence of excessive H^+, but is instead caused by changes in other elements when the system becomes more acidic (e.g., pH decreases). Examples include increased solubilization of metal ions that can be toxic to plants and animals such as Al^{3+} and some trace elements (e.g., Pb^{2+}), more rapid losses of basic cations (e.g., Ca^{2+}, Mg^{2+}), and creation of unfavorable soil environments for soil microorganisms important in many nutrient cycles. In urban or industrial settings, increases in acidity can dissolve carbonates (e.g., limestone, marble) in buildings and other structures and corrode metals in many settings, resulting in considerable aesthetic and economic damage. One long-term urban concern is the effect of acidic deposition on municipal water systems that rely on surface waters to provide drinking water. Many municipalities make extensive use of lead and copper piping in water distribution systems. For instance, questions have recently been raised about the effects of slow dissolution of some metals (Pb, Cu, Zn) from older plumbing materials, due to exposure to more acidic waters, on human health.

10.2.3 Effects of Acidic Deposition on Soils

Soils become acidic by a number of different processes. In humid regions the natural process of weathering acidifies soils because precipitation exceeds evapotranspiration. As mentioned earlier, "natural" rainfall is acidic (pH of ~5.6) and continuously adds a weak acid (H_2CO_3) to soils. This acidification results in a gradual leaching of basic cations (Ca^{2+}, Mg^{2+}) from the uppermost soil horizons, leaving Al^{3+} as the dominant exchangeable cation. Exchangeable Al^{3+} is in equilibrium with

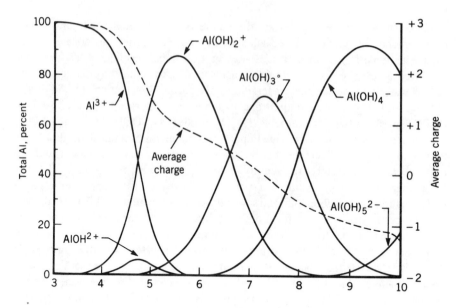

Figure 10-8 Distribution of soluble forms of aluminum as a function of soil pH. (Adapted from Marion et al., 1976. With permission.)

soluble Al^{3+} in the soil solution that can react with water to produce H^+ and thus acidify the soil, as shown below and in Figure 10-8:

$$Al^{3+} + H_2O \rightleftharpoons Al(OH)^{2+} + H^+ \tag{10-2}$$

$$Al(OH)^{2+} + H_2O \rightleftharpoons Al(OH)_2^+ + H^+ \tag{10-3}$$

$$Al(OH)_2^+ + H_2O \rightleftharpoons Al(OH)_3^0 + H^+ \tag{10-4}$$

Other natural processes that contribute to soil acidification include mineralization and nitrification of organic N (see Chapter 4), plant and microbial respiration that produces CO_2 and thus H_2CO_3, and oxidation of FeS_2 in soils disturbed by mining or drainage. However, much of the acidity in soils between pH 4.0 and 7.5 is due to the hydrolysis of Al^{3+}. In extremely acidic soils (pH <4.0), strong acids such as H_2SO_4 are a major component of soil acidity.

Acidic deposition adds strong acids to soils. Under certain conditions this can result in the leaching of basic cations and an increased solubility of Al^{3+}. However, the inorganic and organic chemistry of this system is highly complex (Figure 10-9); and despite an extensive research effort, considerable scientific debate remains as to the specific mechanisms involved and the situations where acidic deposition will have a significant environmental impact. In a practical sense, acidic deposition will have a greater effect on forest soils than agricultural or urban soils because in the latter situations we routinely and inexpensively counteract the effects of all acidifying processes by liming. Although it is possible to lime forest soils, the logistics and cost often preclude this as a routine management practice except in areas severely

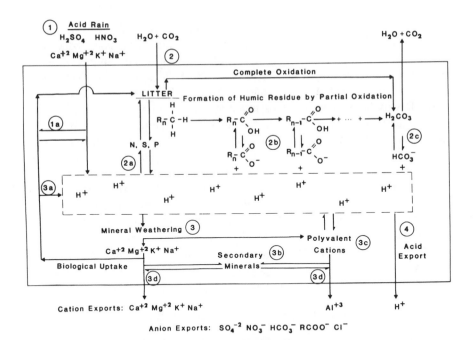

Figure 10-9 Effects of acidic deposition on soils. (1, 1a) Inputs of acid rain add H^+, acidic anions, and basic cations; (2) biological processes such as mineralization (2a) of C, N, S, P produce H^+ but also form weak organic acids that can consume H^+ (2b) or ultimately result in loss of H_2O and CO_2 from soil following complete oxidation (2c); (3) weathering of primary minerals produces cations for plant uptake (3a), secondary minerals (3b), acidic (3c), and basic (3d) cations; (4) export of positive charge from soils as leachable cations (H^+, Al^{3+}, Ca^{2+}, Mg^{2+}, K^+, Na^+) must be balanced electrically by anion export (SO_4^{2-}, NO_3^-, Cl^-, HCO_3^-, organic acids). (Adapted from Krug and Frink, 1983.)

impacted by acidic deposition. The rate and extent of increased acidification beyond that caused by naturally acidic rainfall depends both on the buffering capacity of the soil and the use of the soil. The issue is complicated by the fact that many of the areas subjected to the greatest amount of acidic deposition are also areas where considerable natural acidification occurs. For example, forested soils in the northeastern U.S. are developed on highly acidic parent material and have also undergone tremendous changes in land use in the past 200 years that have contributed to natural acidification. Clear-cutting and burning by the first Europeans to settle in the area has been almost completely reversed and many areas are now totally forested. Humus that accumulates in these forests as they expand represents a natural source of acidity and pH buffering that must be balanced with anthropogenic sources of acidity (e.g., acidic deposition). Similarly, greater leaching or depletion of basic cations by plant uptake in increasingly reforested areas must consider the significant inputs of these same cations in precipitation. Many studies have examined the changes in chemical properties of soils following long-term acidic deposition and changing land-use patterns. While it is beyond the scope of this chapter to discuss this research in detail, several excellent reviews are cited in the References section of this chapter.

Excessively acidic soils are undesirable for several reasons. Direct phytotoxicity from soluble Al^{3+} or Mn^{2+} can occur and seriously injure plant roots, reduce plant growth, and increase plant susceptibility to pathogens. The relationship between Al^{3+} toxicity and soil pH is complicated by the fact that in certain situations organic matter can form complexes with Al^{3+} that reduce its harmful effects on plants. Acid soils are usually less fertile because they lack important basic cations such as K^+, Ca^{2+}, and Mg^{2+}. Leguminous plants may fix less N_2 under very acidic conditions because of reduced *rhizobial* activity and greater adsorption of Mo (as the MoO_4^- anion), a key component of nitrogenase and important enzyme in the N fixation process, by clays and Al and Fe oxides. Mineralization of N, P, and S can also be reduced because of the lower metabolic activity of bacteria at low pH values. It should be noted, however, that many plants and microorganisms adapt to or even prefer very acidic conditions (e.g., pH <5.0). Examples include ornamentals such as azaleas and rhododendrons and food crops such as cassava, tea, blueberries, and potatoes. In fact, considerable efforts in plant breeding and biotechnology are directed toward developing Al- and Mn-tolerant plants that can survive in bioregions such as the tropics where highly acidic soils are commonplace.

10.2.4 Effects of Acidic Deposition on Vegetation

Perhaps the most publicized and visible issue related to acidic deposition has been its effects on forest vegetation. Widespread forest decline has been reported in areas where significant acidic deposition has occurred. In Europe, it has been estimated that as much as 35% of all forests has been affected by acidic deposition. Similarly, in the U.S. many important forest ranges such as the Adirondacks of New York, the Green Mountains of Vermont, and the Great Smoky Mountains in North Carolina have experienced sustained decreases in tree growth for several decades and show serious visible damage. However, as noted earlier, conclusive evidence that forest decline or dieback is caused solely be acidic deposition is lacking and complicated by the many known interactions between acidification and other environmental or biotic factors that influence tree growth. Here we provide a brief summary of some means by which acidic deposition can directly affect plants; indirect effects on plant growth due to soil acidification can also be significant and were summarized in Section 10.2.3.

Wet and dry acidic deposition on leaves may enter directly via the stomata. If the deposition is sufficiently acidic (pH ~3.0), damage can also occur to the waxy cuticle on the surface of leaves, increasing the potential for direct injury of exposed leaf mesophyll cells. Foliar lesions are one of the most common symptoms of plants subjected to simulated acidic precipitation. Gaseous compounds such as SO_2 and SO_3 present in acidic mists or fogs can enter leaves via the stomata, form H_2SO_4 on reaction with H_2O in the cytoplasm of leaf cells, and disrupt many pH-dependent metabolic processes. Many studies have shown increased necrosis of leaves when exposed to high levels of SO_2 gas. The exact mechanisms by which acidification of leaf cells causes injury are not known and certainly vary among species. Physiological changes associated with exposure to simulated acidic deposition include collapsed epidermal cells, eroded cuticles, loss of chloroplast integrity and decreased chlorophyll content, loosening of fibers in cell walls and reduced cell membrane integrity, and changes in osmotic potential that cause a decrease in cell turgor.

Considerable research has been directed at the secondary effects on vegetation exposed to acidic deposition, specifically the likelihood of increased disease or insect damage. The general hypothesis is that any injury to a leaf surface will promote the survivability of pathogenic organisms and their entry into the plant. Studies have shown that leaves exposed to acidic precipitation are more "wettable," a key condition for the establishment of pathogenic populations. Lesions on acidified leaf surfaces are openings that permit ready entry of plant pathogens, in much the same manner as natural wounds. Other effects of acidic deposition on leaves may also predispose them to disease or insect damage. Leaching of nutrients and organic compounds and disruption of photosynthesis and nutrient metabolism weaken the plant's resistance to any type of stress.

Root diseases may also be increased in excessively acidic soils. In addition to the damages caused by exposure to H_2SO_4 and HNO_3, roots can be directly injured or have their growth rate impaired by increased concentrations of soluble Al^{3+} and Mn^{2+} in the rhizosphere. Root exudates (organic compounds secreted from root cells) have been shown to increase when aboveground plant metabolism is affected by acidic deposition on leaves and stems. Changes in the amount and composition of these exudates can then alter the activity and population diversity of soilborne pathogens. The general tendency associated with increased root exudation is a enhancement in microbial populations due to an additional supply of carbon (energy). Chronic acidification can also alter nutrient availability and uptake patterns, and thus aboveground plant growth and yield.

10.2.5 Effects of Acidic Deposition on Aquatic Ecosystems

Ecological damage to freshwaters from acidic deposition is a serious and increasingly well-documented global problem. As with forests, the weight of scientific evidence suggests that a number of interrelated factors associated with acidic deposition are responsible for many of the undesirable changes occurring in aquatic ecosystems. Acidification of freshwaters is, however, not a new phenomenon. Studies of lake sediments suggest that increased acidification began in the mid-1800s, although the process has clearly accelerated since the 1940s. Intensive scientific research on this subject has been conducted since the early 1970s, resulting in hundreds of technical papers and dozens of comprehensive books on the subject (see reference section, this chapter). The subject is complex, and the magnitude of the effects of acidic deposition on fresh waters has been shown to vary widely between geographic regions. This section reviews the major factors that control the extent of freshwater acidification, the most common ecological effects, and some approaches to remediate the problem.

Geology, soil properties, and land use are the main determinants of the effect of acidic deposition on aquatic chemistry and biota, as shown in Table 10-5 for two European watersheds. Lakes and streams located in areas with calcareous geology (e.g., limestone) resist acidification more than those where granite and gneiss are the predominant geologic materials. Soils developed from calcareous parent materials

Table 10-5 Chemical Properties of Precipitation, Throughfall, Leachate, and Stream Waters in Two European Watersheds

Watershed description	pH	H^+	Ca^{2+}	Total SO_4^{2-}	Al (μg/L) Total	Al (μg/L) Soluble
Loch Chon (Scotland)						
Subject to high acidic deposition; rainwater partially neutralized by vegetation and in soil profile; leachate strongly influenced by calcium-rich geology; streams support fish						
		(meq/m²/yr)				
Input						
Above canopy	—	48	—	132	—	—
Below canopy	—	24	—	138	—	—
		(μeq/L)				
Concentration						
Precipitation	4.5	29	11	53	—	—
Throughfall	4.8	15	50	92	—	—
Leachate from O horizon	4.2	68	—	122	—	—
Leachate from BC horizon	4.4	37	—	89	—	—
Stream water	5.1	8	72	93	129	47
Kelty (Scotland)						
Large inputs of sulfate and acidic deposition; spruce canopy contributes to acidification; high sulfate uptake by trees, but little buffering of acidity by soils and geologic material; many streams cannot support fish						
		(meq/m²/yr)				
Input						
Above canopy	—	72	—	140	—	—
Below canopy	—	80	—	230	—	—
		(μeq/L)				
Concentration						
Precipitation	4.5	29	11	53	—	—
Throughfall	4.2	57	45	164	—	—
Leachate from O horizon	4.0	100	—	115	—	—
Leachate from BC horizon	4.1	80	—	73	—	—
Stream water	4.5	34	48	93	140	62

Source: Mason, 1992.

also tend to be deeper and more buffered against acidification than the thin, acidic soils common to granitic areas. Land management also affects freshwater acidity. Forested watersheds tend to contribute more acidity than those dominated by meadows, pastures, and agronomic crops. Trees and other vegetation in forests have been shown to "scavenge" (retain on leaves and stems) acidic compounds in fogs, mists, and atmospheric particulate matter. These acidic compounds are later delivered to forest soils when rainfall leaches them from the surfaces of the vegetation. Rainfall below forest canopies ("throughfall") is often more acidic than ambient precipitation. Silvicultural operations that disturb soils in forests as part of planting, fertilizing, draining, and harvesting trees can increase acidity by stimulating the oxidization of organic N and S, and FeS_2. Conversely, runoff and leachate from watersheds dominated by well-limed agricultural soils can act to neutralize acidity in lakes and streams. However, if agricultural soils are fertilized with ammoniacal N fertilizers and not limed properly, they can also contribute to freshwater acidification. For example, 360 kg/ha of lime (as $CaCO_3$) would be needed to neutralize the acidity from 100 kg N/ha as urea; only 50 kg/ha would be needed to neutralize the acidity of 100 cm of rain with a pH of 4.0. Other factors that can influence freshwater acidification include rainfall intensity and duration and topography and hydrogeology, all of which act to determine the direction and rate of water flow (and thus acidic compounds) through soils and parent material to fresh waters.

A number of ecological problems arise when fresh waters are acidified below pH 5.0, and particularly below pH 4.0. Decreases in biodiversity (number of different species present) and primary productivity (actual numbers and biomass) of phytoplankton, zooplankton, and benthic invertebrates commonly occur. Decreased rates of biological decomposition of organic matter have occasionally been reported, which can then lead to a reduced supply of nutrients as mineralization slows. Microbial communities may also change, with fungi predominating over bacteria. Proposed mechanisms to explain these ecological changes center around physiological stresses caused by exposure of biota to higher concentrations of Al^{3+}, Mn^{2+}, and H^+ and lower amounts of available Ca^{2+}. One specific mechanism suggested involves the disruption of ion uptake and the ability of plants to regulate Na^+, K^+, and Ca^{2+} export and import from cells.

Acidic deposition can also affect fish, amphibians, and mammals. Widespread evidence exists for declining fish populations in acidified lakes and, under conditions of extreme acidity, of fish kills. In general, if the water pH remains above 5.0, few problems are observed; from 4.0 to 5.0 many fish are affected; and below 3.5 few fish can survive. The major direct cause is the direct toxic effect of Al^{3+} which interferes with the role Ca^{2+} plays in maintaining gill permeability and thus respiration. Aluminum toxicity can be a serious problem and can result in major fish kills if any climatic condition, such as heavy rains or rapid snowmelt, substantially accelerates water flow to a stream or lake. Calcium has been shown to mitigate the effects of Al^{3+}, but in many acidic lakes the Ca^{2+} levels are inadequate to overcome Al^{3+} toxicity. Low pH values also disrupt the Na^+ status of blood plasma in fish. Under very acidic conditions H^+ influx into gill membrane cells both stimulates excessive efflux of Na^+ and reduces influx of Na^+ into the cells from the external waters. Excessive loss of Na^+ can cause mortality in fish. Other indirect effects of acidification on fish survival include reduced rates of reproduction, high rates of mortality early in life or in reproductive phases of adults, and migration of adults away from acidic areas. Amphibians are affected in much the same manner as fish, although they are somewhat less sensitive to Al^{3+} toxicity. Birds and small mammals often have lower populations and lower reproductive rates in areas adjacent to acidified fresh waters. This may be due to a shortage of food due to smaller fish and insect populations or to physiological stresses caused by consuming organisms with high Al^{3+} concentrations.

10.2.6 Reversing the Effects of Acidic Deposition

The environmental damage caused by acidic deposition will be difficult and extremely expensive to correct. In the long term, reversal of these effects can only be accomplished by reducing emissions of SO_2 and NO_x. Some approaches to do this include burning less fossil fuel; using cleaner energy sources (e.g., low-sulfur coal or wind and solar systems); and designing more efficient "scrubbers" to reduce the amount of these gases emitted by utilities, industries, and vehicles. There is a general consensus among the scientific community that reducing emissions will result in slow, but eventual improvement in acidified ecosystems, particularly fresh waters. For the present, despite the firm conviction of most nations to reduce acidic deposition, it appears that the staggering costs of such actions will delay implementation of this approach for many years. The U.S. Congress estimated that to reduce SO_2 emissions alone by 9 million metric tons would cost as much as $3.6 billion over a 5-year period; the U.S. Environmental Protection Agency estimated that 20-year costs could be as much as $33 billion.

Short-term remedial actions for acidic deposition are available and have been successful in some ecosystems. Liming of lakes and some forests has been practiced in European counties for over 50 years and is now done in many other areas of the world. Hundreds of Swedish and Norwegian lakes have been successfully limed in the past 20 years. The effectiveness of liming depends mainly on the residence time of water in the lake. Lakes with short water retention times may need annual or biannual liming; others may need to be limed every 5–10 years. As an example, 10 small Adirondack lakes were limed in 1983–1984, and the percent survival of brook trout increased from about 10% at pH 4.5 to 60–80% at pH 5.5–6.5 by 1986; however, three were sufficiently reacidified to cause significant declines in fish populations. The logistics of liming lakes and forests are formidable. Aerial application of lime via planes and helicopters is often required; hence application costs often far exceed the expense of the liming materials. Liming may have some negative effects as well. Much of the vegetation in forested areas is well adapted to acidic soils; liming (or overliming) may alter the distribution of species in these ecosystems in an unpredictable and perhaps undesirable manner.

REFERENCES

Bird, E. and Strange, M., Mares' Tails and Mackeral Scales, Center for Rural Affairs, Walthill, NE, 1992.

Burke, L. M. and Lashof, D. A., Greenhouse gas emissions related to agricultural and land-use practices, Impact of carbon dioxide, trace gases, and climate change on global agriculture, in Kimball, B. A., Ed., ASA Special Publication No. 53, American Society of Agronomy, Madison, WI, 1990.

Charles, D. F., Ed., *Acidic Deposition and Aquatic Ecosytems,* Springer-Verlag, New York, 1991.

Council on Environmental Quality, *Environmental Quality,* 18th and 19th annual report, U.S. Government Printing Office, Washington, D.C., 1988.

Earthquest, Atmospheric Trace Gases That Are Radiatively Active and of Significance to Global Change, Vol. 4, No. 2, Office for Interdisciplinary Earth Studies, Boulder, CO, 1990, 10.

Krug, E. C. and Frink, C. R., Effects of acid rain on soil and water, Bulletin 811, Connecticut Agricultural Experimental Station, New Haven, CT, 1983.

Lashof, D. A. and Tirpak, D. A., Policy Options for Stabilizing Global Climate, Draft report to Congress, U.S. Environmental Protection Agency, Office of Policy, Planning, and Evaluation, Washington, D.C., 1989.

Marion, G. M., Hendricks, D. M., Dutt, G. R., and Fuller, W. H., Aluminum and silica solubility in soils, *Soil Sci.*, 121, 76, 1976.

Mason, B. J., *Acid Rain: Its Causes and Effects on Inland Waters.* Oxford University Press, New York, 1992.

MacCracken, M. C. and Luther, F. M., Ed., Projecting the climatic effects of increasing carbon dioxide (1985), DOE/ER-0237, U.S. Department of Energy, Washington D.C., 1985.

Moore, B., Presentation to the Global Warming Round Table, sponsored by the U.S. Department of Energy, Washington D.C., 1988.

U.S. Environmental Protection Agency, Acid Rain, EPA-600-79-036, U.S. Government Printing Office, Washington, D.C., 1980.

U.S. Environmental Protection Agency, Environmental Progress and Challenges: EPAs Update, EPA-230-07-88-033, U.S. Government Printing Office, Washington, D.C., 1988.

White, R. M., The great climate debate, *Sci. Am.,* 263, 36, 1990.

World Resources Institute, *World Resources, 1988–1989.* Basic Books, New York, 1988.

Wuebbles, D. J., and Edwards, J., *Primer on Greenhouse Gases,* Lewis Publishers, Chelsea, MI, 1991.

SUPPLEMENTARY READING

Conservation Foundation, *State of the Environment: A View Toward the Nineties,* Conservation Foundation, Washington, D.C., 1987.

Kamari, J., *Impact Models to Assess Regional Acidification,* Kluwer Academic Publishers, London, 1990.

Kennedy, I. R., *Acid Soil and Acid Rain,* John Wiley & Sons, New York, 1992.

Linthurst, R. A., *Direct and Indirect Effects of Acidic Deposition on Vegetation,* Butterworth Publishers, Stoneham, MA, 1984.

Reuss, J. O. and Johnson, D. W., *Acid Deposition and the Acidification of Soils and Waters,* Springer-Verlag, New York, 1986.

11   RISK ASSESSMENT

11.1 INTRODUCTION

Risk is defined as the probability or chance of injury, loss, or damage. This is the most general definition of risk that includes items ranging from financial losses or gains, to storm damage to buildings, to human health effects from exposure to pollutants. There is risk to humans and other organisms associated with the concentrations of pollutants and other substances in soils. This chapter summarizes the basic concepts used in risk assessment and illustrates how the risk assessment process is used in writing environmental regulations with an emphasis on soils.

Our definition of risk includes effects on any organism, although the tendency is to focus the discussion on human health effects. Morbidity and mortality, both immediate and delayed, are the two general effects of concern. Risk to organisms is typically expressed as the number of negative outcomes (injury, loss, or damage) divided by the number of organisms exposed to the risk. Other units defining exposure are often associated with risk. For example, an increase in the cancer rate of one case per million people might be after a lifetime exposure to the cancer-causing agent, or traffic fatalities might be tied in with the number of miles driven. Table 11-1 provides risk information for a number of causes of death. The annual individual risk was calculated by dividing the number of deaths per year for each cause of death in 1988 by the population of the U.S. (approximately 250,000,000). This assumes the entire population is exposed to each risk which, of course, is not always true. In other words, the *exposed population* is a subset of the *total population.* Not everyone drives or rides in automobiles, even fewer travel by commercial aviation, and these individuals are not exposed to those risks. The assumption is more valid for causes such as electrocution, accidental falls, or diabetes mellitus where everyone is exposed to the risk to some extent. Nevertheless, such simple calculations allow people to appreciate the relative magnitude of the risk of death by various causes.

There are a number of responses that the general public can make to risks. The risk can be avoided or eliminated, as has been the case for chlorofluorocarbons used as propellants in aerosol cans or dichlorodiphenyltrichloroethane (DDT), where substitutes were available. The cause of the risk can be regulated or modified to reduce the frequency or magnitude of the negative outcome. Building flood control structures or limiting the amount of a substance (e.g., N or trace elements) that could be applied

Table 11-1   Individual Risk for Various Causes of Death on an Annual Basis

Cause of death	Annual individual risk	Ratio[a]
Diseases of the heart	3.11×10^{-3}	321
Cancer	1.97×10^{-3}	507
Motor vehicle traffic accidents	1.92×10^{-4}	5,206
Diabetes mellitus	1.65×10^{-4}	6,070
Suicide	1.22×10^{-4}	8,222
Homicide (by firearm)	5.46×10^{-5}	18,322
Accidental falls	4.84×10^{-5}	20,668
Drowning	1.68×10^{-5}	59,538
Foreign objects or food in respiratory tract	1.52×10^{-5}	65,703
Surgical or medical procedures	1.01×10^{-5}	98,853
Electrocution	2.86×10^{-6}	350,140
Commercial aviation accidents	1.93×10^{-6}	513,347
Lightning	3.28×10^{-7}	3,048,780

Note: Based on 1988 data.
[a] One death per number of people in population.

to soils would be examples of this. The vulnerability of the exposed population can be reduced. In this case the cause of the risk is not changed, but the people potentially affected by it may receive advance warning and be able to reduce their losses. Several postevent strategies are often used. Better ambulance service can be implemented in response to traffic accidents. This increases the chance for surviving an accident but does not change the chance of being in an accident. Insurance also provides financial reimbursement for financial losses.

Risk assessment is a process by which we attempt to determine the probability and magnitude of injury, loss, or damage that may result from a health hazard. *Risk management* is a process by which economic, political, legal, and ethical ramifications of the results of risk assessment are considered. Regulatory decision making is based on both risk assessment and risk management.

There are two major reasons why we would be concerned with risk assessment for soil contaminants. First, if the concentration of a substance in soil is deliberately increased, will any organism experience an unacceptable increase in risk? Second, what increased risks are realized by organisms because of soil contamination that has already occurred? Recall from Figure 3-1 that there are both direct and indirect ways for organisms to be exposed to harmful substances in soils. Therefore, an increase in risk can be realized from the ingestion of soil itself (e.g., by children or by grazing livestock), or indirectly from the consumption of groundwater contaminated by a substance leached from the overlying soil or from the consumption of crops grown in contaminated soil.

11.2 RISK PERCEPTION

A discussion of risk assessment would not be complete without consideration of risk perception by the general public as compared to the scientific community. As you might suspect, the perception of risk by these two groups is quite different and responds to different factors. Surveys have revealed that people generally feel that they face more risks today than people faced in the past. This is despite the fact that

Table 11-2   Over- and Underestimated
Frequency of Death as Judged by Lay People

Overestimated	Underestimated
Tornadoes	Asthma
Floods	Stomach cancer
Pregnancy	Diabetes
Botulism	Stroke

Source: Slovik et al., 1979.

average life expectancy has increased in recent times. Let us consider risk perception by the general public.

It is well established that lay people overestimate the frequency of rare causes of death and underestimate the frequency of more common causes of death, as illustrated in Table 11-2. Far fewer people die as a result of tornadoes, floods, pregnancy, or botulism than were estimated by lay people. The actual number of deaths each year from any of these causes is less than 1000. Likewise, far more people die from asthma, stomach cancer, diabetes, and strokes than were estimated. More than 100,000 people die from strokes each year, for example. This information indicates that familiarity with the cause of death induces a bias in our perception of the risk. A death as a result of botulism is quite unusual and would likely be reported by the news media. Since people are not familiar with botulism as a cause of death, the noteworthy status of the death would raise their consciousness of the cause and could make them perceive that botulism is more common than it really is. A death as a result of a stroke would not be a newsworthy item, and over time people's awareness of the magnitude of this cause of death would diminish.

Other factors that influence risk perception by the general public are outlined in Table 11-3. The two primary factors involved in risk perception by the general public are: *dread* and *knowledge* (whether the risk is known or unknown). Dread is an intense fear of something that might happen. An accident or activity that invokes dread and is relatively unknown will be perceived as the most risky, whereas an accident or activity that does not invoke dread and is familiar will be perceived as the least risky. Obviously, many accidents or activities fall somewhere in the middle in that they may invoke dread, but are familiar to us or vice versa. Whether the risk is undertaken voluntarily is also an important consideration. A person smoking a cigarette is at much greater risk from tobacco smoke than nonsmokers subjected to passive smoke, yet the nonsmokers will have a greater objection to the risk than will the smoker. The classification scheme shown in Table 11-3 identifies many of the important considerations when risk is perceived by lay people and allows one to qualitatively predict how the public will perceive a new risk.

An additional factor that helps explain variations in risk perception by the public over time is that reports of accidents or events serve as signals. This simply means that the public's awareness of a risk, however small that risk may be, is greatly increased by a major event or accident. The Three Mile Island and Chernobyl nuclear accidents made the public's perception of the risks from nuclear energy increase dramatically and have played a major role in slowing or stopping the increased use of nuclear energy in the U.S. Concerns over contamination of Times Beach in Missouri by dioxin heightened the public's awareness of the dangers of dioxin; and

Table 11-3 Factors Associated with Risk Perception by the General Public

Risk perception factor	Subfactors for high or low risk perception		Examples for high or low risk perception within factors	
	High	Low	High	Low
Dread	Uncontrollable Globally catastrophic Fatal consequences Not equitable Involuntary	Controllable Not globally catastrophic Consequences not fatal Equitable Voluntary	Nuclear reactor accident Nerve gas accidents Nuclear weapons (war)	Caffeine consumption Aspirin consumption Power mowers
Knowledge	Effects not observable Unknown to those exposed Effect delayed New risk	Observable effects Known to those exposed Effects immediate Old risk	Electric fields DNA technology Nitrogen fertilizers	Auto accidents Alcohol accidents Handguns

Source: Slovic, 1987.

the discovery of two poisoned grapes in a shipment from Chile raised concerns about food safety in general. Technological advances present a dichotomy of sorts for risk perception. Such advances can increase the perceived risk because they are not understood by the general public, while at the same time the public often relies on technological advances as a salvation from risk.

The public's perception of risk places regulatory agencies in somewhat of a quandary. While these are public agencies designed to respond to public needs and concerns, they cannot justify the allocation of scarce resources to problems that are perceived to be much worse than they actually are.

11.3 CARCINOGENICITY

The risk assessment procedures that are followed are influenced by whether the substance has been shown to be carcinogenic. The U.S. Environmental Protection Agency has a classification scheme for carcinogenicity based on human and animal evidence. This scheme is outlined in Table 11-4. Substances in Group A are known human carcinogens (e.g., radon, vinyl chloride), Group B refers to probable human carcinogens, Group C refers to a possible human carcinogen, Group D refers to something that is unclassified because of inadequate data, and Group E refers to a substance with evidence of noncarcinogenicity. For regulatory purposes Groups A and B are collectively called Category I, Group C is called Category II, and Groups D and E are collectively called Category III.

As an example of how the classification may influence the regulatory process one can use the maximum contaminant level goals (MCLG) for drinking water. The MCLG is the desired maximum concentration for the substance in drinking water considering all of the potential harmful effects from the substance. The MCLG for any Category I substance is zero while those for Category II and III substances are nonzero values calculated in various ways. A concentration of zero is not obtainable, and the actual drinking water standard is called the maximum contaminant level (MCL). The regulations require that the MCL be as close to the MCLG as possible with the best available technology for removing that substance from the water. Obviously, this process tends to force the MCL to be lower for Category I substances than for Category II or III substances.

11.4 RISK ASSESSMENT

The general process of risk assessment consists of one or more of the following steps: *hazard identification, exposure assessment, dose/response assessment,* and *risk characterization.* Hazard identification is a qualitative assessment of a substance which indicates that exposure to that substance may cause harm. One way for a substance to be found harmful would be through simple epidemiological studies in which the frequency of disease was found to be higher in an exposed group compared to an unexposed group. To a limited extent, all new substances are considered to be a hazard until proved otherwise. The combination of exposure assessment,

Table 11-4   Carcinogenic Categorization of Substances Based on Animal and Human Data

Human evidence	Animal evidence				
	Sufficient	Limited	Inadequate	No data	No evidence
Sufficient	A	A	A	A	A
Limited	B1	B1	B1	B1	B1
Inadequate	B2	C	D	D	D
No data	B2	C	D	D	E
No evidence	B2	C	D	D	E

Note: Group A is a known human carcinogen, Group B1 is a probable human carcinogen based on limited evidence of carcinogenicity from human epidemiological
studies, Group B2 is a probable human carcinogen based on sufficient evidence
of carcinogenicity from animal studies, Group C is a possible human carcinogen,
Group D is not classified due to inadequate data, and Group E has evidence of
noncarcinogenicity.

dose/response assessment, and risk characterization are sometimes referred to as
quantitative risk assessment.

11.4.1 Exposure Assessment

Exposure assessment is the process by which the identity of the organisms
exposed to a contaminant and all possible means of exposure to a contaminant are
determined. The relative contribution of each route of exposure to the dose of the
recipient is also investigated. The *dose* is the amount of the contaminant ingested or
inhaled. One needs to consider how the organism can receive a dose as a starting
point. Humans and animals, for example, can be dosed via inhalation, by dermal
exposure, or by ingestion (Figure 3-1). Therefore, air quality, chance of dermal
contact, and possibility of the pollutant entering the digestive tract need to be
considered in exposure assessment. Plants have analogous pathways for exposure.
Plants extract nutrients, contaminants, and water from the soil and are sensitive to
changes in soil composition. Plants respire and respond to changes in air quality.
Substances can also be absorbed on the waxy surfaces of leaves, similar to dermal
exposure in humans; and dustfall and quality of the soil and water contacting the
leaves need to be considered.

An excellent example of exposure assessment relative to potential soil contamination was done for the sludge disposal regulations approved by the U.S. Environmental Protection Agency in December 1992. These regulations have far-reaching
implications. Of interest to this discussion are the limits for the total amount of trace
elements that can be applied to soils via land application of sludge. Table 11-5 lists
all the exposure pathways considered in the risk assessment process for these
regulations. While these pathways were used specifically to consider land application
of sludge, in a general sense they could be appropriate for anything contained in or
applied to soils (e.g., pesticides, manures, fertilizers). The number or types of
pathways would vary, of course, depending on the fate and transport mechanisms of
the pollutant. Each pathway has an organism as an endpoint, and the potential
negative effects on the organism are the concern.

One of the steps in exposure assessment for the sludge regulations was to identify
the most sensitive pathway for each trace element. The most sensitive pathway refers

Table 11-5   Pathway Models for Land Application of Sewage Sludge

Pathway[a]	Description of highly exposed individual (HEI)
1: Sludge→soil→plant→human	General food chain transfer; 2.5% of all plant derived foods produced on sludge-amended soil, lifetime exposure
1F: Sludge→soil→plant→human	Farmland converted to residential home garden use 5 yr after reaching max. sludge application, 50% of garden foods, lifetime exposure
2F: Sludge→soil→human child	Farmland converted to residential use 5 yr after reaching max. sludge application with children ingesting soil (200 mg/d)
3: Sludge→soil→plant→animal→human	Farm households, 40% of meat produced on sludge-amended soil, lifetime exposure
4S: Sludge→animal→human	Farm households, 40% of meat produced on sludge-sprayed pastures, lifetime exposure
4M: Sludge→soil→animal→human	Farm households, 40% of meat produced on sludge-amended soil, lifetime exposure, animals ingesting soil while grazing
5: Sludge→soil→plant→animal	Livestock ingesting food or feed crops grown on sludge-amended soil
6S: Sludge→animal	Grazing livestock on sludge-sprayed pastures, 1.5% sludge in diet
6M: Sludge→soil→animal	Grazing livestock, 2.5% sludge-soil mixture in diet
7: Sludge→soil→plant	Crops grown on sludge-amended soil
8: Sludge→soil→soil biota	Soil biota living in sludge-amended soil
9: Sludge→soil→soil biota→predator	Shrews or birds, 33% of diet is earthworms from sludge-amended soil
9D: Sludge→soil→(soil biota)→predator	Shrews or birds, habitat is sludge-amended soil
10: Sludge→soil→airborne dust→human	Tractor operator exposed to dust
11: Sludge→soil→surface water→human	Water quality criteria
12A: Sludge→soil→air→human	Farm households breathing fumes from any volatile contaminants in sludge
12W: Sludge→soil→groundwater→human	Farm wells supply 100% of water used for lifetime

Source: U.S. EPA, 1989; Ryan and Chaney, 1993; and Chaney and Ryan, 1993.
[a] F = Future, S = surface applied, D = direct, M = mixed with plow layer, A = air, W = water.

to the pathway where an adverse effect would occur at the lowest soil contaminant concentration. Examples include: pathway 7 for B, Zn, Cu, and Ni where phytotoxicities limit plant growth before negative effects on other organisms are thought to occur; pathway 2 for Pb and F where direct consumption of soil causes the greatest problems; pathway 5 for Mo and Se where ruminant animals suffer molybdenosis or selenosis because of plant uptake of these elements occurs without the plants themselves experiencing phytotoxicities or before other organisms are at risk; and pathway 1 for Cd where food chain transfer to humans is the greatest concern. The fate and transport mechanisms for these trace elements in soils clearly play a role in determining which pathway ends up being the most sensitive. For example, Pb is strongly sorbed by soils to the extent that plant uptake and consequently food chain transfer to humans or animals is of little concern, but direct consumption of soil is. Cadmium, on the other hand, is readily taken up by plants and is easily moved from the soil into the food chain.

In conjunction with determination of the most sensitive pathway is the definition of the highly exposed individual (HEI), those persons with exposure greater than

95% of the population and with the most likelihood of suffering greatest harm at the lowest dose of the pollutant. The HEI represents a subdivision of a group of organisms. The basis for the division might be gender, age, cigarette smoking, dietary habits, sources of food, or area of residence. The HEIs for each pathway used for the sludge disposal regulations are also given in Table 11-5. Other examples of HEIs include infants, for nitrates in drinking water; or predatory birds or people with a diet comprised mainly of fish, for DDT. The premise behind the HEI concept is that if regulations protect the most exposed and sensitive individual in the most sensitive pathway, then all other individuals will be protected as well. In practice, a HEI is not used for all risk assessment calculations. Where the HEI is used, however, the selection of an appropriate HEI is important. It is easy to construct a scenario for a HEI that is so restrictive that there are no individuals within the population that fit the description of the HEI. When this occurs, the calculations are for a risk that cannot exist and the results are overly protective.

In order to determine dietary exposure for humans, an accurate determination of what is consumed by people must be made. This is accomplished with a total diet study. The total diet study incorporates a survey of consumers that determined how much of various food groups were being consumed by various gender and age categories of the population. If the composition of one or more of the food groups changes, then the effect of that change on the population can be estimated. Data are also available on average water consumption by adults and children. A value of 2 L per day is often assumed for adults, although this is likely an overestimate for most individuals. Overestimating factors such as water or food consumption is another way to build safety into the risk assessment process. Soil ingestion by children has also been extensively studied. Most children (age 1–6) will consume less than 0.1 g of soil per day with the contribution from indoor dust versus outside soil ranging from 0 to 100%. Some children with pica (children who ingest nonfood items), on the other hand, have been shown to ingest in excess of 8 g of soil per day averaged over 10 days of measurement.

11.4.2 Dose-Response Assessment

In a general sense, dose-response assessment establishes the relationship between the amount of a substance that an organisms receives (the dose) and the effects on that organism (the response). Responses can be favorable or unfavorable. For the risk assessment process, dose-response assessment may take information from the exposure assessment and determine the effects on the exposed organisms.

The dose-response curve forms the conceptual basis for dose-response assessment. These curves can have several general shapes, as shown in Figure 11-1. Figure 11-1(a) indicates no response to the dose, Figure 11-1(b) is a linear response, Figure 11-1(c) is a threshold response, and Figure 11-1(d) is an asymptotic response. Note that the curves do not have to pass through the origin, indicating that the response may occur in the population without exposure to the substance in question. In other words, other factors may induce the response in addition to the substance being studied.

A value that is sometimes calculated from dose-response experiments in the *no observable adverse effect level* (NOAEL). The NOAEL is the highest dose that can

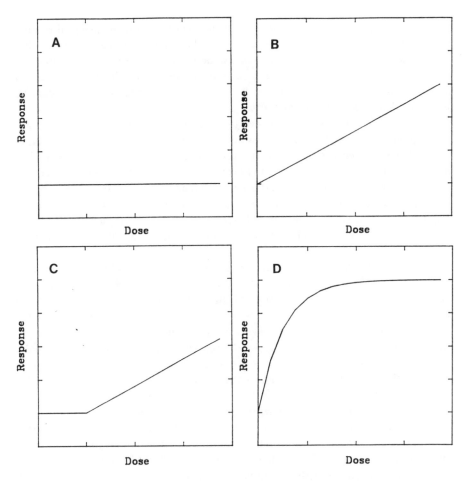

Figure 11-1 Four general shapes for dose-response curves. Curve (a) reflects no response, curve (b) is a linear response, curve (c) is a threshold response, and curve (d) is an asymptotic response.

be administered without a statistically significant increase in adverse response. A linear dose-response curve, for example, predicts an increase in response with each incremental increase in dose, no matter how small the increment is. The increase in response when the dose is increased from zero, however, must become large enough to be statistically significant. The NOAEL is a measure of how much the dose must be increased before a response occurs or before a threshold is surpassed. The NOAEL value is used in some risk assessment calculations.

As precise as the definition for the NOAEL is, few dose-response experiments actually use doses near the NOAEL value such that the NOAEL can be determined with a great deal of confidence. In fact, the doses used in animal experiments are typically much higher than humans are expected to receive. The primary reason for this is that the increase in response seen at the low doses expected in humans is so small that it would take an enormous number of laboratory animals exposed to the low doses over extended periods of time to obtain statistically significant results.

Thus, a major problem in dose-response assessment is the extrapolation of experimental results obtained from high doses given to laboratory rats to the low doses that will actually be realized by humans. Several models may be used for this extrapolation. In general, a linear dose-response curve is assumed for carcinogenic substances and a threshold dose-response curve is assumed for noncarcinogenic substances, unless evidence exists to the contrary.

The extrapolation of results obtained from animal studies to humans is a second major problem in dose-response assessment. It is an accepted belief that substances known to cause cancer in humans will cause cancer in animals, but the opposite has never been conclusively shown. There are obvious moral questions here in that dose-response experiments cannot be performed with human subjects. In addition, epidemiological studies with human populations have limited value in identifying quantitative dose-response relationships. Thus, the use of laboratory animals for dose-response assessment, with all the apparent shortcomings, is a necessary component of risk assessment.

One interesting approach to extrapolating from high dose to low dose and from animals to humans is the Ames test. The Ames test is named after its creator, Bruce Ames, a biochemist at the University of California at Berkeley. The Ames test is a means of ranking possible carcinogenic hazards in humans by calculating a human exposure/rodent potency (HERP) value. The HERP value is calculated by expressing the average daily intake of a suspected carcinogen (mg/kg/day) as a percentage of the TD_{50} for the substance in the most sensitive rodent species. The TD_{50} is the daily dose of the substance that produced tumors in 50% of the lab animals at the end of a standard lifetime. Example HERP values are given in Table 11-6.

The Ames test has been somewhat controversial and is not used in regulatory decision making. It does provide some interesting comparisons, however. The relative risk from some environmental pollutants (e.g., trichloroethylene, PCBs) or pesticide residues (e.g., DDT, ethylene dibromide [EDB]) is less than that from certain pharmaceuticals or from natural carcinogens that are in our food and drink. We also see that the HERP values for two prescription drugs that are commonly taken over extended periods of time are relatively high, as are the values for worker's exposure to formaldehyde or EDB. From the discussion of risk perception, it is apparent that the perceived risk from dietary exposure to PCBs, DDT, or EDB would be much greater than that for the consumption of peanut butter, chlorinated tap water, or alcoholic beverages while the Ames test would indicate otherwise for the real risks based on carcinogenicity.

11.4.3 Risk Characterization

Risk characterization combines the results from exposure assessment and dose-response assessment and determines the management practices that produce acceptable exposures to the receptor organisms. It is often a back calculation, starting with an acceptable exposure to an organism and working back to the management practices or media concentrations (e.g., soil or water concentrations) that produce the maximum acceptable exposures. Such calculations may form the basis for regulatory action.

Table 11-6   Human Exposure/Rodent Potency Index (HERP) Values for
Selected Substances

Daily human exposure	Carcinogen dose per 70 kg person	HERP value (%)
Chlorinated tap water, 1 L	Chloroform, 83 µg	0.001
Contaminated well water, 1 L	Trichloroethylene, 2800 µg	0.004
Chlorinated pool water, 1 hr (child)	Chloroform, 250 µg	0.008
Conventional home air (14 hr/d)	Formaldehyde, 598 µg	0.6
Polychlorinated biphenyls (PCBs), daily dietary intake	PCBs, 0.2 µg	0.0002
Dichlorodiphenyldichloroethylene (DDE), DDT, daily dietary intake	DDE, 2.2 µg	0.0003
Ethylene dibromide (EDB), daily dietary intake	EDB, 0.42 µg	0.0004
Bacon, cooked (100 g)	Dimethylnitrosamine, 0.3 µg	0.003
Peanut butter, one sandwich (32 g)	Aflatoxin, 64 ng	0.03
Mushroom, one raw (15 g)	Mixture of hydrazines	0.1
Beer, 354 mL (12 oz)	Ethyl alcohol, 18 mL	2.8
Wine, 250 mL (8.5 oz)	Ethyl alcohol, 30 mL	4.7
Phenobarbital, one sleeping pill	Phenobarbital, 60 mg	16
Clofibrate (average daily dose, for reducing cholesterol)	Clofibrate, 2000 mg	17
Formaldehyde, worker's average daily intake	Formaldehyde, 6.1 mg	5.8
EDB, worker's average daily intake (high exposure)	EDB, 150 mg	140

Source: Ames et al., 1987.

A value that is sometimes used in risk characterization is the reference dose (RfD). The RfD is the daily intake of a chemical which, if taken during an entire lifetime, will be without appreciable risk. The units of the RfD are typically milligrams per kilogram (mg/kg) body weight per day. The RfD is sometimes calculated as the NOAEL divided by a safety factor on the order of 10–1000, as is the case when animal data are being extrapolated to humans. The safety factor adjusts for comparisons between species and for studies using chronic, subchronic, or acute exposures. Several examples might best illustrate applications of the risk characterization process.

Atrazine is a herbicide commonly used to control weeds in row crops. One question that has been debated is whether atrazine should be classified as a Group B2 or as a Group C carcinogen. One study with one strain of rat found an increased incidence of mammary tumors after exposure to atrazine, but the U.S. Environmental Protection Agency did not feel this was sufficient evidence to place atrazine in Group B2. Several studies determined NOAEL values ranging from 0.48 to 5.0 mg/kg/day. The responses monitored included discrete myocardial degeneration in dogs to second generation lower pup weights in rats. The U.S. Environmental Protection Agency will typically use the lowest NOAEL for the calculation of the RfD. In this case, the RfD was calculated as 0.0048 mg/kg/day (the NOAEL divided by a safety factor of 100). From the RfD, the drinking water equivalent level (DWEL) was calculated as follows:

$$DWEL = \frac{RfD \times BW}{I_w} \qquad (11\text{-}1)$$

where

> BW = body weight of an adult (70 kg typically used)
> I_w = drinking water consumed by the person in a day (2L/day)

The DWEL value is the maximum allowed drinking water atrazine concentration assuming the person obtains all of his/her atrazine from the water. In this case the DWEL is 0.168 mg/L.

The next step is to calculate the MCLG. The MCLG is calculated one of two ways. The preferred way is to adjust the DWEL based on the amount of the chemical obtained from other sources, such as food and air. If this information is not available (the usual case), then the MCLG is calculated assuming that 20% of the chemical is obtained from the drinking water. An additional safety factor of 10 is used for atrazine since it is a Group C carcinogen. This results in an MCLG of:

$$\frac{0.168 \text{ mg/L} \times 0.2}{10} = 0.003 \text{ mg/L} \tag{11-2}$$

The best available technologies allow the MCL to be the same as the MCLG for atrazine. Thus the drinking water standard became 0.003 mg/L (3 ppb), and any public drinking water supplier whose average atrazine concentration exceeds 3 ppb must take steps to reduce it. Atrazine can be removed from water with charcoal filtering, although the process is expensive.

A considerable amount of debate has taken place over atrazine. Some parties feel that sufficient evidence exists to ban atrazine altogether. Atrazine users feel that it is an important tool (effective and inexpensive) in crop production and that atrazine is safe when used properly. The Federal, Insecticide, Fungicide and Rodenticide Act (FIFRA) is the legislation that covers the labeling of atrazine. It stipulates that atrazine should perform its intended function without unreasonable adverse effects on the environment. Unreasonable adverse effects are described as any unreasonable risk to man or the environment, taking into account the economic, social, and environmental costs and benefits of the use of atrazine. Clearly, the law provides a need to balance risks or costs against benefits. The weight of evidence is on the side of benefits for atrazine for the moment. The cost issue is important in this case because of the high cost of charcoal filtering. Public water supplies that use a high proportion of surface water as a source may have problems meeting the 3 ppb atrazine standard if atrazine is used extensively in the drainage area of the water source. In one unique case, a pesticide management area was established in Kansas around a reservoir that places more stringent rules on the use of atrazine than required by the product label. Compliance is voluntary at this point; but without a reduction in the atrazine concentration in the reservoir, they may become mandatory.

A second example of risk characterization pertains to the allowable cumulative application rate of a contaminant to soil via land application of sewage sludge, as had been described previously. The reference application rate (RP, kg/ha) is calculated as follows:

$$RP = \frac{\left[\dfrac{RfD \times BW}{RE} - TBI \right] \times 10^3}{\Sigma UC_i \times DC_i \times FC_i} \tag{11-3}$$

where

RP, BW, and RfD	= as defined before
TBI	= total background intake rate of the contaminant (mg/day)
RE	= relative effectiveness of ingestion exposure (unitless, generally assumed to be 1 unless evidence exists to the contrary)
10^3	= conversion factor (μg/mg)
UC_i	= uptake response slope for the food group i (μg/g dry weight per kilogram of contaminant applied per hectare)
DC_i	= daily dietary consumption of the food group i (gram dry weight per day)
FC_i	= fraction of food group i assumed to originate from sludge amended soil (unitless, assumed to be 2.5% for the population as a whole)

The value of DC_i comes from the total diet study. While the actual calculations are cumbersome and will not be repeated here; however, the reader should appreciate several aspects of Equation 11-3. The equation is written with a human as the receptor organism, although it is appropriate for any organism provided the appropriate data are available. The numerator determines the allowable effective dose of the contaminant to the receptor organism above the background level. The denominator determines the increase in the contaminant concentration in each food group as the contaminant is added to the soil, and factors in how much of that food group is consumed and the proportion of that food group that originates from soils impacted by the contaminant. In the case of direct soil ingestion, the denominator is changed to reflect the change in soil contaminant concentration as the contaminant is added to the soil and the soil ingestion rate. Additional safety factors may be added.

Recall from Chapter 7 that trace element contamination of soils is essentially permanent. So it must be recognized that the sludge disposal regulations allow a certain amount of irreversible soil contamination to occur. This is in contrast to the western European philosophy which argues that soils have a finite capacity to absorb pollutants without negative consequences and that man should not use up any of that capacity (i.e., soil concentrations should not be allowed to change). The U.S. regulations were written with a risk versus benefit philosophy. The risk assessment process indicates the increase in risk can be regulated to insignificant levels, and there are considerable benefits for land applying sludge. Land application of sludge can save money compared to other disposal options. For example, placing the material in a landfill has become prohibitively expensive in recent years and has associated risks such as leachate generation and potential for contamination of surface waters and groundwater. Incineration or ocean dumping are no longer considered environmentally acceptable. The sludge also contains plant nutrients and organic matter that can benefit soil physical properties and soil fertility. Finally, national regulations might actually reduce risk in areas where sludge disposal was not previously regulated.

11.5 UNCERTAINTY

Both the general public and the scientific community can be critical of the risk assessment process and the resulting regulations. The general public often wants unrealistic levels of safety and, at times, seems to be unaware of the fact that a risk-free society cannot exist. The desired levels of risk are much lower than the risks realized from most everyday activities. Such reasoning led to the passage of the Delaney Clause which prohibits the use of cancer-causing agents as food or feed additives. The overall concept of the Delaney Clause is sound except that it provides an absolute risk standard that does not take into account probability, costs, or benefits. The general public is also uncomfortable with arbitrary safety factors that it may not perceive as safe enough. In addition, increases in risk are often stated in terms of a few additional negative outcomes per 100,000 or 1,000,000 people exposed. This leaves the impression that preventable negative outcomes are deliberately allowed to occur.

The scientific community is critical of regulations that are written with limited or suspect data. The previously mentioned problems of extrapolation from high dose to low dose and from animals to humans are the cause of considerable debate. Both extrapolations require making predictions beyond the range of the data, which is generally not acceptable within the scientific method. Scientists always want more data before they are comfortable in making a decision.

Criticism of the risk assessment process is likely to continue indefinitely. It is helpful for both the general public and the scientific community to understand that the risk assessment process is not in quest of scientific truth. The goals are to estimate risk such that it is unlikely that the actual risk is underestimated and to develop regulations that will protect the environment without unreasonable risk expense. Ultimately, environmental regulations and their implications for acceptable levels of risk are a societal, not a scientific decision.

REFERENCES

Ames, B. N., Magaw, R., and Gold, L. S., Ranking possible carcinogenic hazards, *Science,* 236, 271, 1987.

Chaney, R. L. and Ryan, J. A., Heavy metals and toxic organic pollutants in MSW-composts: research results on phytoavailability, bioavailability, fate, etc., in *Science and Engineering of Composting: Design, Environmental, Microbiological, and Utilization Aspects,* Hoitink, H. A. J. and Keener, H. M., Ed., Renaissance Publishers, Worthington, OH, 1993.

Ryan, J. A. and Chaney, R. L., Regulation of municipal sewage sludge under the Clean Water Act Section 503: a model for exposure and risk assessment for MSW-compost, in *Science and Engineering of Composting: Design, Environmental, Microbiological, and Utilization Aspects,* Hoitink, H. A. J. and Keener, H. M., Ed., Renaissance Publishers, Worthington, OH, 1993.

Slovik, P., Perception of risk, *Science,* 236, 280, 1987.

Slovik, P., Fischhoff, B., and Lichtenstein, S., Rating the risks, *Environment,* 21, 14, 1979.

U.S. Environmental Protection Agency, Development of Risk Assessment Methodology for Land Application and Distribution and Marketing of Municipal Sludge, EPA/600/6-89/001, 1989.

SUPPLEMENTARY READING

Glickman, T. S. and Gough, M., Eds., *Reading on Risk, Resources for the Future,* Johns Hopkins University Press, Baltimore, MD, 1990.

Hallenback, W. H. and Cunningham, K. M., *Quantitative Risk Assessment for Environmental and Occupational Health,* Lewis Publishers, Chelsea, MI, 1986.

McColl, R. S., Ed., *Environmental Health Risks: Assessment and Management,* University of Waterloo Press, Waterloo, Ontario, Canada, 1987.

U.S. Environmental Protection Agency, The Risk Assessment Guidelines of 1986, EPA/600/8-87/045, 1987.

Table A-1 The Elements and Their Symbols, Atomic Numbers, and Atomic Weights Based on the Assigned Relative Atomic Mass of $^{12}C = 12$

Name	Symbol	Atomic number	Atomic weight
Actinium	Ac	89	227.028
Aluminum	Al	13	26.98154
Americium	Am	95	(243)[a]
Antimony	Sb	51	121.75
Argon	Ar	18	39.948
Arsenic	As	33	74.9216
Astatine	At	85	(210)
Barium	Ba	56	137.33
Berkelium	Bk	97	(247)
Beryllium	Be	4	9.01218
Bismuth	Bi	83	208.9804
Boron	B	5	10.81
Bromine	Br	35	79.904
Cadmium	Cd	48	112.41
Calcium	Ca	20	40.08
Californium	Cf	98	(251)
Carbon	C	6	12.011
Cerium	Ce	58	140.12
Cesium	Cs	55	132.9054
Chlorine	Cl	17	35.453
Chromium	Cr	24	51.996
Cobalt	Co	27	58.9332
Copper	Cu	29	63.546
Curium	Cm	96	(247)
Dysprosium	Dy	66	162.50
Einsteinium	Es	99	(254)
Erbium	Er	68	167.26
Europium	Eu	63	151.96
Fermium	Fm	100	(257)
Fluorine	F	9	18.998403
Francium	Fr	87	(223)
Gadolinium	Gd	64	157.25
Gallium	Ga	31	69.72
Germanium	Ge	32	72.59
Gold	Au	79	196.9665
Hafnium	Hf	72	178.49
Helium	He	2	4.00260
Holmium	Ho	67	164.9304
Hydrogen	H	1	1.0079
Indium	In	49	114.82
Iodine	I	53	126.9045
Iridium	Ir	77	192.22
Iron	Fe	26	55.847
Krypton	Kr	36	83.80
Lanthanum	La	57	138.9055
Lawrencium	Lr	103	(260)
Lead	Pb	82	207.2
Lithium	Li	3	6.941
Lutetium	Lu	71	174.967 ± 0.003
Magnesium	Mg	12	24.305
Manganese	Mn	25	54.9380

Table A-1 (continued)

Name	Symbol	Atomic number	Atomic weight
Mendelevium	Md	101	(257)
Mercury	Hg	80	200.59
Molybdenum	Mo	42	95.94
Neodymium	Nd	60	144.24
Neon	Ne	10	20.179
Neptunium	Np	93	237.0482
Nickel	Ni	28	58.70
Niobium	Nb	41	92.9064
Nitrogen	N	7	14.0067
Nobelium	No	102	(259)
Osmium	Os	76	190.2
Oxygen	O	8	15.9994
Palladium	Pd	46	106.4
Phosphorus	P	15	30.97376
Platinum	Pt	78	195.09
Plutonium	Pu	94	(244)
Polonium	Po	84	(209)
Potassium	K	19	39.0983
Praseodymium	Pr	59	140.9077
Promethium	Pm	61	(145)
Protactinium	Pa	91	231.0359
Radium	Ra	88	226.0254
Radon	Rn	86	(222)
Rhenium	Re	75	186.2
Rhodium	Rh	45	102.9055
Rubidium	Rb	37	85.4678
Ruthenium	Ru	44	101.07
Samarium	Sm	62	150.4
Scandium	Sc	21	44.9559
Selenium	Se	34	78.96
Silicon	Si	14	28.0855
Silver	Ag	47	107.868
Sodium	Na	11	22.98977
Strontium	Sr	38	87.62
Sulfur	S	16	32.06
Tantalum	Ta	73	180.9479
Technetium	Tc	43	(97)
Tellurium	Te	52	127.60
Terbium	Tb	65	158.9254
Thallium	Tl	81	204.37
Thorium	Th	90	232.0381
Thulium	Tm	69	168.9342
Tin	Sn	50	118.69
Titanium	Ti	22	47.90
Tungsten	W	74	183.85
Uranium	U	92	238.029
Vanadium	V	23	50.9415
Xenon	Xe	54	131.30
Ytterbium	Yb	70	173.04
Yttrium	Y	39	88.9059
Zinc	Zn	30	65.38
Zirconium	Zr	40	91.22

[a] Man-made element or one that occurs only in minute quantities in nature.

Table A-2   Conversion Factors for SI and Non-SI Units

To convert Column 1 into Column 2, multiply by	Column 1 SI Unit	Column 2 non-SI Unit	To convert Column 2 into Column 1 multiply by
Length			
0.621	kilometer, km (10^3 m)	mile, mi	1.609
1.094	meter, m	yard, yd	0.914
3.28	meter, m	foot, ft	0.304
1.0	micrometer, μm (10^{-6} m)	micron, μ	1.0
3.94×10^{-2}	millimeter, mm (10^{-3} m)	inch, in.	25.4
10	nanometer, nm (10^{-9} m)	Angstrom, Å	0.1
Area			
2.47	hectare, ha	acre	0.405
247	square kilometer, km^2 (10^3 m)^2	acre	4.05×10^{-3}
0.386	square kilometer, km^2 (10^3 m)^2	square mile, mi^2	2.590
2.47×10^{-4}	square meter, m^2	acere	4.05×10^3
10.76	square meter, m^2	square foot, ft^2	9.29×10^{-2}
1.55×10^{-3}	square millimeter, mm^2 (10^{-3} m)^2	square inch, in^2	645
Volume			
9.73×10^{-3}	cubic meter, m^3	acre-inch	102.8
35.3	cubic meter, m^3	cubic foot, ft^3	2.83×10^{-2}
6.10×10^4	cubic meter, m^3	cubic inch, in^3	1.64×10^{-5}
2.84×10^{-2}	liter, L (10^{-3} m^3)	bushel, bu	35.24
1.057	liter, L (10^{-3} m^3)	quart (liquid), qt	0.946
3.53×10^{-2}	liter, L (10^{-3} m^3)	cubic foot, ft^3	28.3
0.265	liter, L (10^{-3} m^3)	gallon	3.78
33.78	liter, L (10^{-3} m^3)	ounce (fluid), oz	2.96×10^{-2}
2.11	liter, L (10^{-3} m^3)	pint (fluid), pt	0.473
Mass			
2.20×10^{-3}	gram, g (10^{-3} kg)	pound, lb	454
3.52×10^{-2}	gram, g (10^{-3} kg)	ounce (avdp), oz	28.4
2.205	kilogram, kg	pound, lb	0.454
0.01	kilogram, kg	quintal (metric), q	100
1.10×10^{-3}	kilogram, kg	ton (2000 lb), ton	907
1.102	megagram, Mg (tonne)	ton (U.S.), ton	0.907
1.102	tonne, t	ton (U.S.), ton	0.907
Yield and Rate			
0.893	kilogram per hectare, kg/ha	pound per acre, lb/acre	1.12
7.77×10^{-2}	kilogram per cubic meter, kg/m^3	pound per bushel, lb/bu	12.87
1.49×10^{-2}	kilogram per hectare, kg/ha	bushel per acre, 60 lb	67.19
1.59×10^{-2}	kilogram per hectare, kg/ha	bushel per acre, 56 lb	62.71
1.86×10^{-2}	kilogram per hectare, kg/ha	bushel per acre, 48 lb	53.75
0.107	liter per hectare, L/ha	gallon per acre	9.35
893	tonnes per hectare, t/ha	pound per acre, lb/acre	1.12×10^{-3}
893	megagram per hectare, Mg/ha	pound per acre, lb/acre	1.12×10^{-3}
0.446	megagram per hectare, Mg/ha	ton (2000 lb) per acre, ton/acre	2.24
2.24	meter per second, m/sec	mile per hour, mi/hr	0.447
Specific Surface			
10	square meter per kilogram, m^2/kg	square centimeter per gram, cm^2/g	0.1
1 000	square meter per kilogram, m^2/kg	square millimeter per gram, mm^2/g	0.001

Table A-2   (continued)

To convert Column 1 into Column 2, multiply by	Column 1 SI Unit	Column 2 non-SI Unit	To convert Column 2 into Column 1 multiply by
		Pressure	
9.90	megapascal, MPa (10^6 Pa)	atmosphere	0.101
10	megapascal, MPa (10^6 Pa)	bar	0.1
1.00	megagram per cubic meter, Mg/m³	gram per cubic centimeter, g/cm³	1.00
2.09×10^{-2}	pascal, Pa	pound per square foot, lb/ft²	47.9
1.45×10^{-4}	pascal, Pa	pound per square inch, lb/in²	6.90×10^3
		Temperature	
1.00 (K − 273)	Kelvin, K	Celsius, °C	1.00 (°C + 273)
(9/5 °C) + 32	Celsius, °C	Fahrenheit, °F	5/9 (°F − 32)
		Energy, Work, Quantity of Heat	
9.52×10^{-4}	joule, J	British thermal unit. Btu	1.05×10^3
0.239	joule, J	calorie, cal	4.19
10^7	joule, J	erg	10^{-7}
0.735	joule, J	foot-pound	1.36
2.387×10^{-5}	joule per square meter, J/m²	calorie per square centimeter (langley)	4.19×10^4
10^5	newton, N	dyne	10^{-5}
1.43×10^{-3}	watt per square meter, W/m²	calorie per square centimeter minute (irradiance), cal/cm² min	698
		Transpiration and Photosynthesis	
3.60×10^{-2}	milligram per square meter second, mg/m² sec	gram per square decimeter hour, g/dm² hr	27.8
5.56×10^{-3}	milligram (H_2O) per square meter second, mg/m² sec	micromole (H_2O) per square centimeter second, μmol/cm² sec	180
10^{-4}	milligram per square meter second, mg/m² sec	milligram per square centimeter second, mg/cm² sec	10^4
35.97	milligram per square meter hour,	milligram per square second, mg/m² sec	2.78×10^{-2} decimeter
		mg/dm²)hr	
		Plane Angle	
57.3	radian, rad	degrees (angle), °	1.75×10^{-2}
	Electrical Conductivity, Electricity, and Magnetism		
10	siemen per meter, S/m	millimho per centimeter, mmho/cm	0.1
10^4	tesla, T	gauss, G	10^{-4}
		Water Measurement	
9.73×10^{-3}	cubic meter, m³	acre-inches, acre-in.	102.8
9.81×10^{-3}	cubic meter per hour, m³/hr	cubic feet per second, ft³/sec	101.9
4.40	cubic meter per hour, m³/hr	U.S. gallons per minute, gal/min	0.227

Table A-2   (continued)

To convert Column 1 into Column 2, multiply by	Column 1 SI Unit	Column 2 non-SI Unit	To convert Column 2 into Column 1 multiply by
8.11	hectare-meters, ha-m	acre-feet, acre-ft	0.123
97.28	hectare-meters, ha-m	acre-inches, acre-in.	1.03×10^{-2}
8.1×10^{-2}	hectare-centimeters, ha-cm	acre-feet, acre-ft	12.33
Concentrations			
1	centimole per kilogram, cmol/kg (ion exchange capacity)	milliequivalents per 100 grams, meq/100 g	1
0.1	gram per kilogram, g/kg	percent, %	10
1	milligram per kilogram, mg/kg	parts per million, ppm	1
Radioactivity			
2.7×10^{-1}	becquerel, Bq	curie, Ci	3.7×10^{10}
2.7×10^{-2}	becquerel per kilogram, Bq/kg	picocurie per gram, pCi/g	37
100	gray, Gy (absorbed dose)	rad, rd	0.01
100	sievert, Sv (equivalent dose)	rem (roentgen equivalent man)	0.01
Plant Nutrient Conversion			
	Elemental	Oxide	
2.29	P	P_2O_3	0.437
1.20	K	K_2O	0.830
1.39	Ca	CaO	0.715
1.66	Mg	MgO	0.602

Source: American Society of Agronomy, Madison, WI. Reprinted with permission.

Table A-3   Journals and Periodicals Related to the Environmental Sciences

Advances in Soil Science
Agricultural Chemistry
Agrochimica
Agronomy Journal
Ambio
American Society of Agronomy Abstracts
Analyst
Analytical Chimica Acta
Analytical Letters
Analytical Chemistry
Atmospheric Environment
Canadian Journal of Fisheries and Aquatic Science
Canadian Journal of Chemistry
Canadian Journal of Soil Science
Communications in Soil Science and Plant Analysis
CRC Critical Review of Environmental Contamination
Crop Science
Environmental Geology and Water Science
Environmental Letters
Environmental Research
Environmental Science and Technology
Environmental Toxicology and Chemistry
Estuarine, Marine and Coastal Shelf Sciences
Geochimica et Cosmochimica Acta
Geoderma
Geological Society of America Bulletin
Journal of Chemical Ecology

Journal of Chemical Education
Journal of Environmental Quality
Journal of Environmental Health
Journal of Atmospheric Chemistry
Journal of Environmental Science and Health
Journal of Environmental Engineering (Division of ASCE)
Journal of the Geological Society (London)
Journal of Soil Science
Journal of the Air Pollution Control Association
Journal of the Association of Official Analytical Chemists
Journal of Environmental Science Technology
Journal of Great Lakes Research
Journal of the Water Pollution Control Federation
Journal of Environmental Toxicology and Contamination
Journal of Marine Research
Journal of the American Water Works Association
Journal of the Indian Chemical Society
Journal of Soil & Water Conservation
Limnology and Oceanography
Marine Chemistry
Marine Environmental Chemistry
Nature
Science
Soil Science
Soil Science Society of America Proceedings
Talanta
Water, Air, and Soil Pollution
Water Research
Water, Science and Technology
Weed Science Journal

INDEX

major, in soil science, 53–54
Environmental quality, introduction to, 1–5;
    see also different aspects of
    environmental quality
Environmental regulatory agencies, nitrogen
    management and, 84
Environmental science, general public and,
    4–5
Environmentalism, overview of, 2
EON rates, see Economically optimum
    nitrogen rates
EPA, see U.S. Environmental Protection
    Agency
EPIC, see Erosion-productivity-impact-
    calculator
Epidemiology, characterization of, 3–4
Epilimnion, phosphorus and, 121, 125
Epipedons, soil classification and, 39–40
Equilibrium phosphorus concentration, at zero
    sorption, 119, 121, 125, 138–139
Equipment costs, nitrogen sources and, 92
Eriophyes chondrillae, biological control of,
    212
Erosion
    characterization of, 73
    global climate change and, 244
    nitrogen and, 58, 61, 64, 71, 73–75, 96
    organic matter and, 30
    pesticides and, 193
    phosphorus and, 106, 108, 123–124,
        135–136
    phytotoxicity and, 173
    runoff and, 18
    sediments and, 50
    soil animals and, 32
    soil development and, 37
    soil environment and, 20
    soil physical properties and, 22
    soil quality and, 52
    sulfur and, 162
    trace elements and, 179
Erosion-productivity-impact-calculator (EPIC),
    phosphorus and, 130, 133
ESP, see Exchangeable sodium percentage
Ester herbicides, effects of, 190
Ester sulfates, in soil, 155
Estuarine systems, phosphorus and, 104
Ethylene dibromide (EDB)
    dose response assessment and, 272–273
Europe, acidic deposition and, 250, 256–258
European corn borer, biological control of,
    211, 213
European spruce sawfly, biological control of,
    211
Eutrandept, phosphorus and, 118

Eutrophication
    biogeochemical cycles and, 227
    definition of, 103
    hydrosphere and, 13–14
    leaching and, 71
    nitrogen and, 56, 59
    phosphorus and, 103–106, 110, 127, 129
    sediments and, 50
    water quality and, 14
Eutrophy
    hydrosphere and, 13–14
    phosphorus and, 103
Evaporation, hydrologic cycle and, 15
Evaporation beds, sulfur and, 156
Evaporites, sulfur and, 147
Evapotranspiration
    acidic deposition and, 253
    hydrologic cycle and, 18
Evesboro soil, phosphorus and, 117, 121
Evolution, theory of, 4
Exapion ulicis, biological control of, 212
Exchangeable sodium percentage (ESP), sodic
    soils and, 163–164
Excluders, trace elements and, 178
Explosions, origination of airborne particles
    from, 10–11
Exposed population, risk assessment and, 263
Exposure assessment, risk and, 268–270
Extraction, organic chemicals and, 207

Facultative anaerobes,soil microorganisms and,
    35
Fallowing, global climate change and, 244
Farms, biogeochemical cycles and, 222–225
Fat, human exposure to soil pollutants and, 52
Federal Insecticide, Fungicide and Rodenticide
    Act (FIFRA), atrazine and, 274
Feedlots
    ammonia volatilization and, 68
    biogeochemical cycles and, 223
    leaching and, 72
    organic chemicals and, 186
    phosphorus and, 111
Feldspar, particle density of, 23
Fermentation wastes, nitrogen and, 82
Ferredoxin, sulfur and, 159
Fertigation, nitrogen and, 93–95, 99
Fertility
    soil quality and, 51–52
    trace elements and, 181
Fertilizers
    ammonia volatilization and, 68
    biogeochemical cycles and, 222–224
    exposure assessment and, 268
    global climate change and, 244

Soils
  composition of, 21
  hydrologic cycle components and, 17–18
Solar radiation, global climate change and,
    237–238
Solidification, organic chemicals and, 207
Solidifying agents, trace elements and, 181
Solid-phase treatment, in windrows,
    bioremediation and, 232
Solid wastes
  application of, 95
  biogeochemical cycles and, 224
  hazardous wastes as, 48
  nitrogen and, 80
Solubility
  organic chemicals and, 194–195
  organic matter and, 30
  pesticides and, 193
Solute potential, hydrologic cycle and, 17
Solvent property, of water, significance of, 15
Solvents, organic chemicals and, 185
Sorghum, global climate change and, 245
Sorption
  organic chemicals and, 198–200
  trace elements and, 181
Soybeans
  global climate change and, 245
  nitrogen fixation and, 77
  phosphorus and, 115, 137
  sulfur and, 161
  zinc and, 183
SPAD meter, in assessing nitrogen-sufficient
    soils, 93
Species diversity, low, water quality and, 14
Sphecidae, biological control and, 213
Spider mites, biological control of, 213
Spiders, biological control and, 213
Spinner spreaders, application of solid organic
    wastes by, 95
Spiny prickly pear, biological control of, 211
Spodosol, 41, 150
Spores, as atmospheric particulate matter, 10
Spotted knapweed, biological control of, 212
Staphylinidae, biological control and, 213
"Starter" fertilizers, phosphorus and, 131
Steam, nitrogen and, 78
Steel manufacturing
  phosphorus and, 113
  trace elements and, 167
Stink bugs, biological control and, 213
St. John's wort, biological control of,
    211–212
Stones, soil physical properties and, 21
Storage, nitrogen sources and, 92
Stratified soils, hydrologic cycle and, 17

Stratopause, pressure and temperature
    variations in, 9
Stratosphere, 8, 9, 56
Straw, immobilization and, 65
Streams, hydrosphere and, 12–13
Stream stripping, organic chemicals and,
    206–207
Strong acids, acidic deposition and, 253
Structure, soil
  characterization of, 22–24
  organic matter and, 29
  soil classification and, 39
  soil physical properties and, 22
Subangular blocky shape, soil physical
    properties and, 22–23
Subgroup, soil classification and, 39
Suborder, soil classification and, 39
Subordinate designation, soil development and,
    37–39
Subsurface flow, phosphorus and, 105, 107,
    118, 127–128
Subsurface horizons
  denitrification and, 72
  hydrologic cycle and, 17
  phosphorus and, 127
  soil classification and, 39–40
  sulfur and, 150
Sugarcane, sulfur and, 161
Sugars, organic matter and, 29
Sulfates
  adsorption and desorption of, 153–154, 156
  forms of inorganic sulfur in environment
    and, 146
  soil physical properties and, 24
  trace elements and, 181
Sulfides
  inorganic sulfur in soils and, 146, 149
  trace elements and, 181
Sulfites, forms of inorganic sulfur in
    environment and, 146
Sulfolobus spp., sulfur and, 157
Sulfur
  adsorption and desorption of soil sulfate,
    153–154, 156
  agriculture and, 161–163
  atmospheric cycles and, 11
  availability indices of, 160–161
  biogeochemical cycling and, 221–224, 230
  biosphere and, 32
  deficiency of, 144, 158–160
  delay of nitrification and, 96
  environmental impact of, 144–146
  global sulfur cycle and, 146–153
  importance of, 143–144
  inland surface waters and, 18